D1117628

Iron Oxen

William Hinton

IRON OXEN

A Documentary of Revolution in Chinese Farming

Vintage Books

A DIVISION OF RANDOM HOUSE/NEW YORK

To the young mechanics, tractor drivers, technicians, and laborers on China's state farms who went wherever they were needed, turned wastelands into fertile fields, and built a socialist way of life.

In a few decades, why can't 600 million "paupers," by their own efforts, create a socialist country, rich and strong? The wealth of society is created by the workers, the peasants, the working intellectuals. If they take their destiny into their own hands, use Marxism-Leninism as their guide, and energetically tackle problems instead of evading them, there is no difficulty in the world which they cannot overcome.

—*Mao Tse-tung*

Contents

Prologue 1

1. Mule Cart to South Ridge 29
2. "I've Never Seen a Freight Train" 34
3. Wasteland Pioneers 42
4. Director Li 47
5. Lao Hei Gets Married 56
6. Donkeys, Rifles, and Tractor Engines 63
7. "I Accept Your Criticism" 69
8. Tractors Need Room to Turn Around 75
9. Battles North and South 84
10. Drama Under a Rising Moon 92
11. Peking on a Tractor 98
12. Liberated Tientsin 107
13. Plowing in Thousand Ching Basin 113
14. Sugar-Coated Bullets 119
15. Trains, Shops, and a Pacifist 125
16. Autumn Sowing 134
17. Militiaman Kuo Passes the *Gate* 141
18. Double Bridge 146

19. Soybeans for Steel   152

20. Wherever the People Need Us   161

21. A Trip to the South   169

22. Big Yang Sits under the Pomegranate Tree   182

23. When the Sun Stands One Stalk High   190

24. Combines in the Cotton Patch   196

25. It Works   200

26. People Come First   206

27. *Pu Chien Tan*   211

    Postscript, 1970   215

Iron Oxen

Prologue

TIEH NIU—iron oxen—is the name peasants gave to tractors when they first appeared on China's far flung plains after the end of World War II.

This is a book about tractors in China and the men and women who first put them to work in the Liberated Areas of the North. There, because land reform had been completed and a new revolutionary government held power, large-scale farming and farm mechanization was at last able to establish a firm base in the land of the ox and the hoe.

In pre-revolutionary China mechanization had been all but impossible. No one seriously considered introducing tractors, grain drills, and combines into a countryside where tillage methods remained as stagnant as the medieval rural society they reflected.*

The United Nations Relief and Rehabilitation Administration (UNRRA), which sent almost 2,000 tractors to China between 1945 and 1947, never planned to transform native agriculture with them. UNRRA machinery was originally destined for the area in Honan flooded by the Yellow River. There millions of acres of good land had been laid waste in 1937 when Chiang Kai-shek ordered the breaching of the dykes to delay Japan's advance southward. The wild grasses that took over the land after the first deluge subsided spread roots so thick that peasant plows could not break them. Relief tractors were meant to be sod-

---

* The virgin lands of the Northeast were a minor exception. There tractors were introduced between the two world wars and some mechanized farming was carried on for many years. Lanz, of Germany, had a dealer in Shenyang (Mukden) who sold and serviced tractors on a commercial basis.

busters that would break up and turn the heavy roots so that returning refugees could begin once more the age-old cycle of winter wheat and summer beans on fields plowed with animal or even human power and harvested with sickles. The tractors, once they had accomplished their task and were worn out from taming several million acres, were to be discarded.

Even the Japanese, who built ten-thousand-acre farms on the salt flats of the Pohai Gulf during World War II, never introduced machinery for cultivation. Huge diesel engines and rotary pumps for irrigation, yes—but the lands reclaimed by power pumping were divided into countless small plots and rented out to Korean tenants familiar with rice production. These tenants did most of their work by hand. They were migrants from poverty-stricken villages in mountainous Korea, dependent on the Japanese for seed, fertilizer, water, and fuel, were no better off than serfs, and had to accept whatever contract terms their overlords saw fit to grant. Why should anyone consider tractors when manpower such as this was available?

As for China's traditional gentry—families who, depending on the region, owned from one-third to three-quarters of all the land in the country—they rarely invested in such fundamentals as improved seed, chemical fertilizer, insecticides, or modern implements, not to mention tractors and combines. They held scattered plots, not large estates, and their bailiffs were specialists in rent collection, not farm management. If these gentry prospered they bought additional land with their surplus grain, converted it into gold and buried it in the ground, or loaned it out at exorbitant interest rates to tenants who had not kept enough food from the harvest to survive the winter.   ·

How could tractors be grafted onto such a system? Who dared risk capital on imported machines when 100 percent a month could be earned on short-term loans to the desperate poor? And even if someone wanted to introduce tractors, where were the lands they could work? Few plots, even on the flat North China plain, equaled half an acre in size, while most of the paths between fields were too narrow for modern machines and implements to pass along.

True, there were wastelands, millions of acres of wastelands, but who dared undertake their reclamation? Certainly not any private citizen and certainly not the corrupt and avaricious bureaucrats of the Kuomintang. They left the wastelands for foreigners to tackle.

Ironically, when UNRRA, hoping to stabilize the status quo, through resettlement, finally sent tractors by the hundreds to reclaim the huge flooded areas of Honan, the realities of Chinese politics prevented any major concentration of machinery there. Provincial governors and military commanders in other parts of the country were not willing to see supplies and equipment worth millions of dollars sent to one spot to enrich a colleague or rival. No: the tractors had to be split as many ways as there were governors and generals with influence. And so, instead of establishing one great reclamation project in Honan, with central repair shops and satellite depots, with a few skilled engineers and many technicians, the tractors, implements, parts, and supplies were split into many shipments and sent, ten or twenty units at a time, to provinces north, south, east, and west, wherever wasteland could be found to serve as an excuse. On reaching their destination, these tractors sometimes plowed land and sometimes sat idle, but always served as a lever for enriching those who controlled them. Imposing bureaucracies of relief-fund recipients were built around these tractors as the men in charge went up a rung on the ladder of wealth and political influence.

I went to China in January 1947 as a technical volunteer with the Church of the Brethren Unit which had recruited most of the instructors for the UNRRA tractor program. In 1947 the Nationalists still breathed confidence in the outcome of the civil war which they were forcing on the country with American support. Their armies were on the offensive everywhere, driving into and splitting up Communist-held areas all over North and Northeast China. Mismanagement of relief and reconstruction projects, however, caused many UNRRA staff members to question whether the Nationalists could ever rebuild China even if they should succeed in conquering it. Public-spirited officials with faith in the future, willing and able to do the hard spadework necessary to establish tractor stations, could not be found. Young

people who wanted to devote their lives to the grueling labor
of mechanized tillage no doubt existed, but, more often than not,
those chosen by local officials as trainees were careerists who had
no intention of soiling their hands with manual labor as a way
of life. Most of the wasteland available either needed far more
than tractor plows to bring it under cultivation or was already
under the control of local gentry who knew well how to pare
the flesh off any project while leaving the bare bones to the poor
peasants and refugees UNRRA had ostensibly set out to help.

My first lesson in these social realities came in the Kuomintang-
controlled areas of the Northeast, where I spent six weeks in
February and March.

The economy of China's foremost industrial region was in a
shambles that winter. The cities looked as if they had been
through several years of bombardment. Row on row of substantial brick houses, once occupied by privileged Japanese, had
been gutted, their windows smashed, their roofs torn off, their
floors torn up. All that remained were brick walls and gaping
holes that had once held doors and window frames. Factories had
been similarly stripped and looted. Mile on mile of smokeless
chimneys stood against the gray winter sky. Everyone blamed
the destruction on the Russians, but inquiry revealed that the
Russians had removed only key items of heavy equipment. The
bulk of the damage had been done by the local population after
the Russians had left and before the Kuomintang troops arrived.
The liberated people had revenged themselves on their imperial
conquerors and forestalled Kuomintang looting by tearing everything of value out of Japanese-built factories and houses. They
mainly sought wood that could be used as fuel and hardware
that could be sold. What they could not carry off, they often
smashed. The destruction was tremendous, but thousands of families lived for many months on the sale of nuts, bolts, gears,
bearings, belts, and other odd items expropriated during those
tempestuous days.

While surveying the prospects for mechanized farming in this
devastated region, we tractor men stayed at the Yamato Hotel
in the heart of downtown Mukden. It was one of those monstrous

buildings that imperialists (in this case Tzarist Russians) were wont to build in all parts of the world, hot or cold. The first floor was two stories high so that having a meal in the dining room was like eating in the grand hall of the old Pennsylvania Station in New York City. Cold drafts, which seemed to originate somewhere above the multilayered glass chandeliers, swept the room periodically and set everyone shivering. In the foyer, colored light from stained glass windows set high in the walls filtered down on a green tile fountain that had long since ceased to flow. Brown woodwork, brown walls, and dark gray ceilings unwashed since Tzar Nicholas' execution added to the gloom.

Yet the Yamato Hotel was booked solid with the most fascinating assortment of people on the make. High Kuomintang generals, their aides and concubines occupied a substantial proportion of the rooms. They shared hallways and lobbies with politicians shorn of their home boroughs and landlords shorn of their homelands. Mayors, governors, and police chiefs appointed to rule in provinces that had not as yet been "liberated" from the grip of the revolutionary hordes exchanged complaints with the worried gentry in whose interest they proposed to govern.

Across the hall from me lived a landlord who claimed twenty thousand acres north of Changchun. Tall, lean, American-educated, polished and yet at the same time a little rough-hewn, this impressive gentleman informed me that General Tu Yu-ming's armies would soon restore the sources of his wealth. Next door lived an Englishman named Edgar who was trying hard to recover the seventeen houses he had once owned in Yingkow. When he had finally managed to reach that port city he had found twelve of his houses completely demolished. When he had gone to the mayor to complain he had been presented with a bill for all the back taxes of the war years. Still he had not given up. He had a childlike faith that the old imperial world would be put back together again and that his own little corner of it would return to his control.

Conversation at mealtimes turned mostly on soybeans, for the Yamato was also headquarters for a motley crew of speculators from Shanghai. Their contribution to world order was to buy soybeans from Manchurian middlemen at two or three cents a pound, bribe the officers in control of transport to get railway cars to ship them south, and then dispose of them at Yellow Sea ports for seven or eight cents a pound, thus realizing fortunes overnight. "There is the man who made $300,000 last month," and "Here comes Li who made $100,000 on one shipment," were typical of the rumors that periodically sent ripples through the dining room.

Although huge shipments of soybeans from the 1946 crop had already been dispatched to France and England, mountains of Manchuria's most famous product still dotted the hinterlands. At every station along the railroad lines we traveled, beans were stacked in bags twenty high, many without cover of any sort and some sitting in pools of water. As the water gradually soaked up through the piles, destroying the value of the beans, traders forced hinterland prices lower while world prices climbed.

A majority of the speculators were actually agents for TV Soong and other "family" companies. "Family" in this case meant the Kungs, Soongs, Chens, and Chiangs who dominated the political, military, and economic affairs of Nationalist China. These soybean buyers were perhaps the least vicious of the "carpetbaggers" sent by those "four" to take over the Northeast in 1945. Most of the generals and top civilian officials ferried by American airforce planes into Manchuria after the Japanese surrender and Soviet pull-out were trusted Kuomintang insiders from the Yangtze Valley. While the Northeast's own great general, the Young Marshall Chang Hsueh-liang, languished in a Nanking villa, still under house arrest for his revolt in Sian in 1936, political hacks who had spent the war enriching themselves in Chungking, Kunming, and Kweichow swarmed into Manchuria as if it were a conquered colony. They seized and shipped off everything of value that had not already been expropriated by the poor. Those few plants and installations that had miraculously survived the transfer of power were taken over more or less intact by southern officials who soon stripped them cleaner than the populace had stripped plants elsewhere. Anything made of metal and in working order brought many times its value in Central China. After a year under Kuomintang care, the economy was in a worse shambles than anyone had conceived possible when its minions arrived.

Under conditions such as these it was clear that there was no bright future for tractors. We found many abandoned industrial buildings that could be converted into storage sheds, repair shops, and tractor-drivers' living quarters. We also found trained mechanics seeking employment. But there was almost no wasteland to plow in the Mukden-Changchun area, the region was already choked with agricultural produce that could not be moved to market, and bureaucrats on the make congregated like bees around honey wherever valuable equipment or relief supplies appeared.

I went with an UNRRA dietician to survey a flooded area where local officials had requested famine relief for 100,000 starving people. The county magistrate put us up overnight in a wine factory which belonged to a large landlord. In the factory yard sat two storage bins made of reed mats. These bins, twelve feet across and at least twelve feet high, were stuffed to the brim, one with soybeans, the other with kaoliang—approximately forty tons to the bin. Other rooms in the compound were also filled with grain, enough to feed thousands. With all this wealth in front of him, the police chief, a large, pot-bellied man with drooping jowls, confided that he was a refugee from the Communists in Hupei Province, that he had twelve family members to support, that his salary was only 8,000 yuan a month, and that he urgently needed relief to carry them through until spring. He urged us to give his case serious consideration.

We did. We found that he commanded no less than 300 armed men, enough to seize all the grain in the district, that he lived alone in the magistrate's compound with no one to support, and that if he was a refugee from Communism it must have been some time ago because all the Communist-led regular forces had long since withdrawn from Hupei.

The Communist threat was the main stock-in-trade of such officials. The mere mention of red hordes was supposed to so unnerve the Western experts that we would immediately unleash the vast stores of supplies which we supposedly controlled. No tale was considered too fantastic. One day Professor McKonkey, a Canadian who was head of all agricultural rehabilitation work in the Northeast and was known to his UNRRA colleagues as the "Emperor of Manchuria," lost his Russian-style great coat in the foyer of the Yamato Hotel. Colonel Sam Li, aide to General Tu Yu-ming, personally investigated the case. After two days of fruitless search he informed us in hushed tones that the coat had been seized by a Communist agent to be used as a disguise for spying at the front.

With equal seriousness Dr. Pan, Minister of Agriculture for the Northeast, insisted that he would have to interview all prospective tractor students personally to make sure there were no Communists among them. He assured us that he could tell a Communist by talking with one. He recommended, to head the whole program, a man who had spent twelve years supervising tractor operations in the field in—of all places—Germany. He had done such good work that he had been awarded a medal by Hitler himself.

After six weeks in the Northeast I was prepared for difficult conditions elsewhere. I was not disappointed. In April I was sent to Suiyuan Province to set up a tractor project in the wastelands east of Paotou, the Yellow River terminus of the Peking-Suiyuan railroad. Our project was located at Salachi, a county town in the middle of an alkaline plain bounded by high rock mountains on the north and the meandering arms of the Yellow River on the south.

From the crest of the mountains on a clear day one could see almost the whole of the Suiyuan plain and out across the Yellow River to the rising lands of the Ordos Desert far beyond. South of the river lay sand; to the north, grass flats, cropland, and a few scattered trees covered the earth with a layer of vegetation too thin to prevent the soil from blowing. When the wind came up great twisters of dust rose in the bed of the river, while here and there across the plain smaller dust squalls whirled, sank, and finally dissipated only to appear again in some other place.

Trees, isolated and small, were hardly noticeable in that vast expanse. Remove them and the appearance of the plain would

hardly be altered. The cropland and grassland, instead of being
a deep green, was a yellow gray just tinted with green. It was as
if a thin veil had been laid on the earth. Looking down between
clumps of grass and stalks of millet to the bare gray earth itself,
it was clear that growing things covered but a small fraction of
the total area of the plain. Plants, animals, and men were accidents
here, scarcely tolerated by nature. One had the feeling that
men were pressing the dry land very close to its limit. The
scattered villages consisted of small clusters of mud houses—adobe
walls, dusty streets, a sprinkling of trees. Each was centered in a
small patch of scrawny crops, wherever lesser concentrations of
salt allowed such crops to grow.

It was a formidable task for any man to support himself in this
country. All winter he shivered in his mud hut while the ground
froze five feet down. Spring came, but it brought no rain. The
drought lasted for weeks and months. Then suddenly, in July, the
climate changed. The sky opened and for hours on end water
poured down. The bare rocks of the mountains could not hold it;
the baked land could not absorb it. Out of the canyons burst
seething brown floods that covered the land and washed out the
crops, smashed the houses and then were gone. In two days it was
as if the rain had never been. The earth was baked as hot and dry
as ever. Peasants digging into it with a stick found that all that
water had penetrated but a few inches. Another year of meager
crops could hardly be avoided.

During the great famine of 1936 relief funds from America had
poured into Suiyuan to build a canal that was to end the awful
cycle of drought and flood east of Paotou. Under the direction
of an American engineer named Todd, a host of foreign super-
visors and thousands of peasants dug a canal forty miles into the
plain from the top of the Yellow River bend. They called it the
Min Sheng or "People's Life" Canal. It was supposed to carry
enough water to irrigate tens of thousands of acres and create a
garden at the foot of the mountains.

Unfortunately, Todd miscalculated and, though water entered
the ditch when the river was at medium height, it was still ten
feet below the land it was supposed to irrigate. When, on the
other hand, the river hit flood stage, the canal served as an aque-

duct to flood the whole countryside and bring up salt that destroyed whatever crops were already under cultivation and permanently damaged the soil. After Todd departed, the canal was renamed the *Min Szu* or "People's Death" Canal.

In 1947 money was again pouring into Suiyuan from abroad in the form of UNRRA flour for dyke repair workers and tractors for the reclamation of wasteland. Officialdom from Kueisui to Paotou buzzed with excited anticipation. It had been a long time since the bureaucrats of this remote area had been able to approach wealth such as this.

Some of the most valuable equipment accompanying the tractors got lost long before it reached Salachi. The packing cases that contained spare parts, a welding outfit, mechanics' tools, and a complete blacksmith shop were broken open in the warehouse on the Tientsin docks and openly looted. When we saw the damaged cases, Jim Wilson, my Brethren Service partner, and I decided to accompany our shipment up the railroad in person. Little did we know what a protracted and hazardous journey we were embarking upon. From Kalgan in Chahar to Tatung in North Shansi all the bridges on a one hundred mile stretch of track had been blown up by guerrillas. Since the bridges, for the most part, crossed dry arroyos, not running streams, maintenance crews had avoided rebuilding them by laying track down the side of each embankment, across the dry watercourse, and back up the embankment on the other side. They had thus converted the railroad into something resembling a slow-motion roller-coaster. The train, traveling at less than five miles an hour, plunged periodically off the right-of-way to negotiate creek beds and was derailed twice in one day. Crews of railway workers exhibited extraordinary ingenuity in coaxing the derailed locomotive (later the tender) back onto the track, but a more serious derailment of the east-bound train forced us to spend the night in the open. Most of the passengers—merchants, speculators, landlords, and military men—showed great anxiety at the prospect. They found an abandoned station house, barricaded themselves inside, and posted guards all around lest bandits or Communists (they used the same word, *tu fei*, for both) raid them in the night.

After five days sleeping on flat cars with our tractors and equipment, we finally reached Salachi and set up shop in the compound of the county farm on the southern outskirts of the town. The compound was large. It needed to be, for we had two officials for each post on the staff and the posts were twice as numerous as the project required. This doubling up was due to rivalry between the regional relief officials sent out from Taiyuan, where the cohorts of General Yen Hsi-shan were supposed to supervise work in three provinces, and the local Suiyuan men, followers of General Fu Tso-yi, who actually commanded the troops on the spot. Since Yen's star was falling (he had lost most of Shansi and large parts of his army, first to the Japanese and later to the Communist-led People's Liberation Army) and Fu's star was rising (he still had a large army, had taken much of Chahar back from the Communists, and was getting supplies direct from the United States), Fu's men had the inside track and assumed all key posts on the farm. But Yen's men could not be sent home and, although there was nothing for them to do, they occupied rooms—not only for themselves, but for their wives and children as well—ate meals, and demanded and received paychecks.

The position of Tractor Captain was assumed by no less an official than the Minister of Agriculture of Suiyuan Province; he brought with him three advisors, all graduates of technical colleges and all absolutely useless, either in the field or the repair shop. The Tractor Captain commanded a Financial Section replete with cashier and bookkeeper; a General Affairs Section that included an executive, several cooks, janitors, room boys, and buyers; a Repair Section that boasted a Japanese engineer and two assistants; and a Drivers' Brigade composed of graduates of the Provincial Agricultural School who had no intention of being tractor hands, but who saw this training as a stepping stone to more exalted careers.

The ten tractors that finally reached Suiyuan were thus supervised by a staff of about thirty, exclusive of drivers, most of whom were rarely seen during the day. Toward evening, however, the compound came to life and the hours between dusk and midnight were gay with talk, song, and the finger game, which served as an excuse for the consumption of large quantities of kaoliang liquor.*

As for wasteland, there was plenty of it around Salachi because of the alkalinity spread by the People's Death Canal, but there was not much point in plowing land on which no crops would grow. Two public farms, one owned by the province and one by the county, contained between them some 16,000 acres, but all that could be tilled had already been rented out to tenants. In order to find land to plow, the *pao changs* (heads of 100 families) and *chia changs* (heads of ten family groups) from the surrounding villages were called into conference. Though they were dressed in the same padded clothes worn by most peasants, many of these men were obviously gentry. Assured that taxes would in no way be affected, they were enthusiastic about the prospect of free plowing and drew up a map showing the tracts to be plowed. Since these were privately owned or came under the control of temples and clans, it seemed unlikely that the poor peasants of Salachi would reap any benefit whatsoever; yet it was the poor peasants and not the gentry we had come to help.

This meeting was held at a time of real crisis for the peasants of Salachi. Yields had been falling and taxes had been rising for several years. Some months earlier, peasants in one village had refused to deliver their tax grain. The magistrate's guard had visited their settlement, arrested the men, stripped their clothes off, poured cold water on their naked bodies, and locked them in an empty room in the below zero weather of a Suiyuan winter. There they froze to death. This object-lesson in law and order had apparently not been brutal enough, however, for not long after we arrived in Salachi several thousand local people went to the railroad station and looted several carloads of wheat and millet awaiting shipment to Tientsin. The Tientsin merchants,

---

* The finger game consists of throwing out the fingers of one hand, from o to 5, and calling out a number which one hopes will equal the total of fingers on one's own and one's opponent's hand. Whoever calls the correct number wins (there can be many draws between each win). The loser has to empty a thimble cup of 90-proof kaoliang distillate.

who had paid large sums of gold for the grain, were outraged. They had been robbed in broad daylight; they demanded compensation and revenge.

The people saw the "outrage" in a different light. Before the merchants had arrived, ordinary townspeople had barely been able to buy the grain they needed at current price levels. After the merchants had bid the prices up, ordinary townspeople could no longer afford to eat. The price escalation was of no help to the average peasant, either. The peasants had long since turned over the bulk of their harvest to landlords and tax collectors, or had sold it to local merchants for a few dollars to buy coal and oil. Now they, like the townspeople, were faced with soaring costs for food and seed. A rumor spread that Magistrate Kao had enriched himself on the deal: he was said to have received two bags of grain for every ten he helped the merchants buy. A mass meeting developed spontaneously in the town square. Many people spoke up and accused the government, the magistrate, and the merchants. They urged action. In the end the people went down to the railroad and took the only action possible—they took back their grain.

This open defiance of authority astonished and angered Magistrate Kao. He was a short man with a round pockmarked face, shaven head, large watery brown eyes, and a row of huge upper teeth, each one of which was set apart from the others and stained brown with nicotine. When he smiled, and he smiled incessantly when entertaining distinguished guests such as tractor technicians from abroad, these stained teeth stood out like a picket fence across his mouth. But in spite of his uncouth appearance, Kao did not seem to be a vicious man. No doubt he would have been content to spend the rest of his days feasting, sleeping, whoring, stamping official papers, and taking kickbacks from Tientsin merchants.

Unfortunately, the authorities in the provincial capital held Kao responsible for everything that happened in Salachi, and most of all for maintaining "law and order." The magistrate wanted to keep his job. He was not a local son. He was not even from Suiyuan. He and his staff had come all the way from

Taiyuan (some remnant of Yen Hsi-shan's influence) and were a long way from home. If Kao should be demoted, they would all be stranded. As a man of honor, fully aware of his responsibility for relatives, friends, and retainers, the magistrate had no choice but to do whatever was necessary to maintain law and order in his domain.

But this time Kao did not know where to begin. The incident had happened so suddenly and so many people had taken part that he did not know whom to hold responsible. Since he couldn't arrest everyone, he sat tight and called for help. A high-ranking general on Fu Tso-yi's staff, a man born and raised in Salachi, came all the way west from Kalgan to investigate. At a mass meeting in the town square he rebuked the assembled townspeople and called them thieves and robbers. What kind of patriotism was this, he cried, to steal grain and loot poor mechants at a time when Fu Tso-yi himself had no time for eating and sleeping, so busy was he defending the country from red bandits. The grain had better be returned, every catty* of it. If it was not, the soldiers would search out every kernel and confiscate it. The guilty would be punished. Let the people not try to take the law into their own hands. Should they ever be dissatisfied with the way things were going in Salachi, let them go to the local magistrate, just, kind, brave Mr. Kao, and let him settle their grievances. So spoke the general and departed.

Still Kao took no action. He couldn't find any ringleaders to arrest. But the provincial authorities sixty miles to the east in Kueisui knew better. The man responsible for this crime could be no other than the wartime guerrilla leader who had, without authorization, broken into Salachi and given battle to the Japanese garrison. They had been wanting to get him for a long time. Had not Salachi's own general proclaimed that if he ever caught the guerrilla hero he would kill him?

Suddenly six truckloads of armed men commanded by the provincial chief of police rolled into town from the east. The ex-guerrilla and several of his friends were quickly arrested. The

*A catty is equal to one-half a kilogram or 1.1 pounds.

people of the town were herded into the central square at rifle point and threatened once more. Then the trucks, the armed men, their captain, and his prisoners left as quickly as they had come, raising a great cloud of dust across the hot plain.

Soon after that eight armed men robbed the cash box at the railroad station and got away unharmed. Next the provincial farm was held up in broad daylight by men with automatic rifles. That same week an employee of the county farm was stripped of all his belongings, including his clothes, just outside the town wall.

Magistrate Kao did the only thing he knew how to do. He tightened control over everything and everybody. He doubled the guard that checked passes at the railroad station. He doubled the sentries at the town gate. He conscripted poor townspeople and peasants by the hundreds to repair and pile higher the town's dilapidated walls, and he sent a contingent under armed guard to throw up earthworks around the railroad station. There was no longer any pretense. Magistrate Kao's county was ruled by naked force.

I left Suiyuan with few regrets in July when UNRRA's North China office recalled me and sent me south across the lines to Communist-held southern Hopei, where twenty tractors awaited a technician to help put them to work. They had been dumped into the Liberated Area as the result of an ultimatum from UNRRA's top leaders that some relief supplies must go to the Japanese-devastated Communist areas or all relief to China would cease. The Brethren Service Committee volunteers who had accompanied the tractors all the way from Shanghai had quit in protest after a quarrel with Liberated Area relief officials and I was sent to salvage something from the debacle. Crossing the battle line, which was the turning point of my life and work in China, was itself unforgettable.

The lone jeep I traveled in churned its way through the endless bog that the monsoon rains had made of the North China plain. The sturdy vehicle, lightly laden with only two people and a few boxes of medical supplies, was covered with layer on layer of mud.

As it plowed forward a spray of silt and water fanned out from each of the four wheels and drummed on the leaves of bean and corn planted tight against the right of way.

It seemed strange that there should be such extensive crops in this place, for there was not a village in sight. The earth was flat as far as the eye could see, and, with the exception of an occasional tree, nothing broke the line of the horizon.

"Never seen it as still as this," said the driver, a stocky White Russian named Igor.

"How come?" I asked.

"No man's land," said Igor. "Too damn quiet!"

He stepped down hard on the accelerator, gripped the wheel with both hands, and concentrated on staying on the high side of the road, which had become nothing but a few cart tracks in the mud. The jeep rushed forward, smashing through water holes and all but leaving the ground wherever the route crossed one of those slight elevations that marked some ancient boundary between fields.

Suddenly a man appeared on the track in front of us. He was running, flying toward us. He had a gun in his hand and brought it to his shoulder as he ran.

"Watch out . . ."

Igor jammed on the brakes with all his strength. I tried to duck my head below the dash but my knees got in the way.

The gun went off.

There was no impact. The bullet went over our heads. The jeep lurched to a stop and stalled three feet from the sentry, who now stood, gun in hand, ready to pull the trigger again.

We were already surrounded by men with rifles. They seemed to have risen out of the earth. I caught a brief glimpse of weapons and clothing—an old rabbit gun with a five-foot barrel, a German Luger in a wooden holster, homespun black tunics held together with cloth fasteners, white towels on shaven heads, a blue visor cap set off with a badge struck in rough likeness to Mao Tse-tung, cloth shoes mud-splattered and soaking wet, soil curling upward under the pressure of bare toes.

These were not soldiers, I realized, but militiamen—the famed *min ping* of the Chinese Revolution. They and they alone stood guard at the border.

The man with the visor cap and the Luger examined Igor's papers, smiled, and waved us on.

"Next time don't come so fast," he called as we pulled away.

Ahead of us a narrow walk topped a wide-open, multi-gated sluice. The jeep bounced across the walk and plunged down into the fertile plain beyond. We had entered the Communist-led Liberated Area. It was like a plunge into a dream world, unreal because it was so quiet and so calm. There was not one soldier, one single trench or any kind of fortification in sight—in contrast to the Nationalist territory we had left behind, where every town and village had been converted into a mud-walled fort. There was no evidence of any military activity in the course of the next thirty miles. Travelers were few and far between: invariably they turned out to be peasants or peddlers lugging carrying-poles, pushing wheelbarrows, or driving mule carts on routine rural errands.

The muddy track, which Igor followed as if by instinct, wound in and out over the small undulations of the plain amid a jungle of growing crops that seemed, under the influence of the monsoon, to have run riot. Kaoliang twelve feet high, corn leaves a full handbreadth across, completely blocked all vistas. Only at the edges of the villages, where threshing floors made small clearings of bare ground, were there any breaks in the double wall of green. Coming suddenly upon such openings, we caught sight of mud-walled dwellings under arbors of trees, women, bare to the waist, spinning cotton in their dooryards, naked children rolling in the gutters. It was as if nothing ever happened, had ever happened, or would ever happen to disturb the pastoral fecundity of this faraway world.

For the next four months I lived and worked in the small county town of Chihsien, which lay on the flat belly of the North China plain at a point almost equi-distant from the encircling ring of the Kuomintang armies. There, behind a tight blockade, one thousand miles from the nearest parts depot, with tractors running on gasoline worth $5 a gallon, seventy students plowed 3,000 acres of sod so that local peasants could plant wheat. My job was to teach the students the care, repair, and maintenance techniques required by modern farm machinery. But far more absorbing than this job was my effort to understand the social revolution that was surging all around me.

It did not take me long to realize that the outward medievalism of rural life in South Hopei concealed currents of profound inner change. If the same kind of wooden-wheeled carts that we had seen on the other side of the lines rattled down the same narrow tracks, if the same hardworking peasants urged their slow oxen across the same fields, if the same wide-eyed, earth-smeared children played on the banks of the village pools, and if the same low, mud-walled houses provided the same shelter for man and beast—still, if one looked closely one saw that this sameness was only superficial. There was, for instance, the matter of clothing. Nowhere did I see anyone expensively dressed, and nowhere did I find anyone in rags, a thing unheard of in traditional China. Most of the clothing was made of homespun, handwoven and hand-dyed cloth. The thread was coarse, the weaving slightly irregular, and the dye fast fading, but the clothes wore well and cost the people only their labor. Since cold fall weather was coming, women were working diligently to prepare the padded suits that would carry their families through the winter. We saw them everywhere, on doorsteps, in courtyards, out under the trees, spinning, weaving, sewing, and fluffing cotton. Some of the clothes they were making were winter uniforms for soldiers. Such self-sufficiency had long since disappeared in Nationalist-held areas.

If the people would be warm that winter, they would also eat well. Observers told me that the land had not been farmed so intensively or vigorously for years. When the summer crops had done poorly the previous year, the peasants had planted wheat on an immense scale. Wheat had always been a gamble— even if one succeeded in getting a crop, the landowners and the government took most of it—but now that the land belonged to the people, they were willing to take their chances with the weather. Never had so high a proportion of the land been sown to wheat.

And never had so little land lain idle. I noticed only scattered patches of uncultivated land. I asked about one of these, a good-looking five *mou** plot near the experiment station at Linching. Mr. Chang, head of the station, told me it belonged to a landlord whose acreage had been expropriated. He had never had to work before and found field labor intolerable. By himself he was un-able to till the land that still was his.

Cotton, too, had been planted more widely than ever before. The revolutionary government had launched a special campaign to interest peasants in growing cotton and, most important, had offered a guaranteed price for it. As a result, half the land in South Hopei seemed to be in cotton.

Such prosperity on the land had stimulated commerce. I had thought that in an area so cut off from the outside world trade would languish. But business in towns like Linching, Weihsien, Hantan, and Nankung was brisk. The shops were full of locally made goods and even of goods from across the lines. In all com-mercial centers, I was told, the number of shops had greatly increased, even doubled, since the days of the Japanese occupa-tion. With land and crops of their own, the peasants had money to spend for the first time in their lives. Merchants and craftsmen found it hard to keep up with demand.

The most striking thing about the towns was the absence of beggars. I cannot say that there were no beggars in the Liberated Areas, for I saw one, a ragged man holding out a bowl on the main street of Wuan. I also saw three blind men wandering about Linching with a little brass gong. They may have been beggars or they may have had some trade. With these exceptions, beg-gars seemed to have completely disappeared. It was unbelievable but it was true. The same went for prostitutes: there did not seem to be any. I was never importuned even though I wandered day and night on the main streets and back alleys of the biggest towns in the area. In Nationalist-held Peking, on the other hand, clerks

* There are six *mou* to the acre.

and roomboys in the main hotels doubled as pimps, while little children touted for their sisters on the sidewalks.

Even more astonishing was the lack of crime. You could drive up to a building in a jeep loaded with baggage, jump out, go inside, and talk for three hours. When you came out there would be a large crowd around the jeep looking everything over, but not one item would be missing. I had boxes full of personal belongings scattered about my room at the tractor school in Chihsien. The door was never locked and the compound gate was never shut, yet I never lost anything. This was in sharp contrast to my experience in Nationalist-held areas, where anything not under lock and key disappeared almost instantly.

We had some valuable tools and spare parts to maintain the tractors and it was some time before we managed to set up any kind of tool check system. But throughout the term of the school, even though large kits of tools were taken daily to the fields, we never lost a single item.

Obviously, law and order was no problem in the Liberated Areas and as a reflection of this soldiers were almost never seen; if one did see a soldier, he was usually on leave. Soldiers on leave carried guns, which was also something new. In Kuomintang China, one rarely saw off-duty soldiers with guns: they could not be trusted to bring the weapons back.

But if soldiers were few and far between, militia were everywhere. They could be seen drilling in the early morning on village threshing floors. They went to town with their rifles slung over their shoulders and thin sacks of millet on their backs—tall peasants dressed in blue or black homespun cotton with white towels on their heads. And they could be seen on their way to joint maneuvers or going to fight in support of regular forces at the front. They moved in great crowds on the highway or rested in the shade by the road, two or three hundred at a time. Here the whole people was armed.

An armed people was evidently in no need of walls. Everywhere town walls had been or were being torn down. At Linching the population had used the city wall to mend the river dyke. Townspeople in Chihsien took thousands of wall bricks home to repair their houses and rebuild what the Japanese had destroyed. When the plowing season ended, we used bricks from the town wall to block up the tractors. In the Liberated Areas ancient fortifications had become a prime source of construction materials for all purposes. With their walls flattened, the towns became a part of the countryside instead of bastions where nervous landlords stored their wealth.

To one who had just come from setting up a tractor project in Kuomintang-held Suiyuan Province, the difference between China Liberated Area Relief Administration (CLARA) officials and Nationalist relief officials was striking. In Kuomintang areas the top man was top man. His word was law; what he wanted was supposed to be done. (Although if his subordinates didn't like it, they sabotaged it in one way or another.) When he was away, all decisions hung fire. No one would take any responsibility. In the Liberated Areas it was different. Here responsible men leaned over backward to get mutual agreement on policy all the way down the line. There were no one-man decisions on important issues, but rather committee decisions. Higher committees showed considerable reluctance to force any action on lower ones. They sought rather to explain and persuade.

Time and again I told old Mr. Tu, head of CLARA's Linching office, that for the successful operation of the tractor project this or that had to be done. "Yes," he would say, "I am convinced you are right. But I cannot just order it done. You see, we must talk it over. Everyone must know the reason why this is essential. We must have a meeting about it and convince them." This was a slow way to work. It could be carried to extremes, but in the long run the results were better. People who did not agree with an order were apt to sabotage it, no matter who the boss might be. The Communists, at least those in CLARA, tried to bring everyone in on the making of decisions and to enlist their support in carrying them out.

On the tractor project itself, democracy was the keynote. Staff and students met regularly to discuss problems and presented criticism to those in charge whenever they felt like it. The students decided their daily schedule—their working hours and their study time. They elected their own foremen and their own squad leaders. Having to stand up to the criticism of the majority and abide by their decisions made the work of a foreign advisor more complicated than it might have been, but it also made it very interesting.

Nearly every day, Mr. Chang, head of the project, would come to me with some new complaint or idea that the students had had—they weren't getting an equal chance to drive, the hours in the field were stretching out so that they had no time to study, I was wasting gasoline, and so on. One of the most frequent requests was that more time be spent on lectures and study pertaining to motors and farm machinery. I was an advocate of learning by practice, and also felt that, from the UNRRA point of view, we ought to plow as much land as possible before the ground froze, and so resisted classroom work as much as I could. A constant good-natured tug-of-war was waged over this issue, which we finally settled with a compromise: I agreed to give lectures on rainy days. As it turned out, it didn't rain from then on, so I won after all.

When it came to using the plowed land, everyone, from the poorest peasant to the county magistrate, had his say. Some of the farmers, those with stock and tools, were all for parcelling out the land to those who could work it. The poor farmers, on the other hand, were for pooling it, village by village, and working it together. A series of meetings was held in the villages to discuss the problem, and village leaders met with sub-district leaders more than once. In the end the poor farmers, who were in the majority, had their way and the land was pooled.

The government men with whom I worked were sincere and hard working. They asked little for themselves. Since I was a foreigner, it was felt that I needed special feeding, and for the first few days the project staff ate good food with me. Then their consciences began to hurt. One morning I came to the table to find it set for only two, the interpreter and myself. The rest had decided that my level of eating was too extravagant for them, too far above the people's diet. And so they banished me to a sumptuous solitude while they ate simpler fare in another part of the compound. I objected strenuously to this isolation and so once again we compromised. We ate simpler food and we ate together.

As a result Chihsien became known to UNRRA personnel as the worst place in the area to stop for a meal.

During the last weeks of the summer drought, I had a chance to see how these men reacted to a crisis. By August, the crops in Chihsien were beginning to show the effects of months of dry weather. The local people were alarmed. They began to haul water to their fields on their backs in the hope of saving at least a few square yards of crop. Old women and small children were struggling over the hot roads with carrying-poles. We had a big water trailer and four large freight trailers. We also had some empty gasoline drums. We decided to put this equipment into service hauling water. But before anything was done, those in charge first found out who needed help the most, whether there were any widows, or sick, or disabled people who were unable to help themselves. Then we hauled water for them.

It was many times impressed on me how hard the government men worked. One day I went to the sub-district office in search of Mr. Yu, head of Chihsien's third sub-district. There was no one there but the clerk. Everyone else was out in the villages on business. We decided we would find Mr. Yu in the countryside and set out after him on a tractor. It turned out to be a wild ride, for the man we sought was on a bicycle and evidently had a lot of people to see that day. Every time we arrived in a village scores of children would rush up to tell us that he had just left for the north, south, east, or west, as the case might be. He was certainly no stranger to them. We finally caught up with him at a small office far out at the county border.

But it was not only the lower civil servants who worked hard, lived simply, and tried to understand the peasants' needs. I met and worked with top government officials of the same caliber. One of these was Mr. Yuan, head of the Civil Affairs Bureau of the Chin-Chi-Lu-Yu Border Region, who was also a CLARA committee man. He was always on the move. He traveled from one end of the region to the other by truck, cart, horse, mule, and even on foot. He looked into conditions everywhere. One rainy night a group of us were stuck in the mud with six vehicles a few miles west of Linching. Rain had been falling for fifteen hours and in places the mud on the road was three feet deep. We could neither advance nor retreat and finally had to sleep on a cold *kang** in a village inn. That very night, Mr. Yuan came through on his horse for an important meeting in Linching. Two days later he left again.

These men were not by any means completely preoccupied with agricultural problems. The development of transportation and industry was also important, and evidence of this was everywhere. We saw long stretches of road built in a few days. In Hantan a large motor pool and a transport depot were under construction. Strong new bricks were being used. The motor pool workers—mechanics, drivers, and office staff—were building their own brick dormitory. The Hantan railroad was already being rebuilt.

The largest project of all was the dyke work on the Yellow River, but this I did not see myself. However, I did see how the government and people met the threat of flood on the Grand Canal with a tremendous mobilization of men and material. In August, as the water in the canal rose, more than 100 miles of dyke were built up several feet in a few days. Weak spots were strengthened and brush reinforcements placed at the curves. Night and day watchers manned lookout stations which were located every few hundred yards. At night one could see the tortuous course of the river by the signal lanterns hung at each station. The water rose to within three inches of the top of the emergency embankments, and then receded. By united effort, a major disaster had been averted and the Grand Canal had been held within its banks. It was only later, after Chiang's troops deliberately cut the dykes south of Tangkuantun, that Central Hopei was flooded.

Even though mobilizing to take action for their own protection and well being was taken for granted by the people in the Liberated Areas, it was a continual source of surprise to visitors. I was once embarrassed, for instance, when I went to confer with the CLARA men in Linching about the difficulty of getting

---

*A *kang* is a sleeping platform of mud brick which has flues from the cooking stove running under it to provide heat.

our plowed land planted without grain drills; I urged immediate action, only to return to Chihsien and find the entire area in question already planted. During the few days that I had been gone, village meetings had been held, seed distributed, plans quickly made, and almost 10,000 *mou* of land planted. The people simply went out with all the animals, seeders, rollers, and labor they had and put in the crop. One group of fourteen villages planted 5,000 *mou* in one morning. To be truthful, I believe that by this effort they even surprised themselves.

Mutual aid groups, where as many as 80 to 100 families pooled labor, stock, and tools to do their work in common, were widespread. But the mobilization of the people did not stop there. Even small children were drawn into social effort whenever possible. Not only were they going to school and helping around the farms, but they were organized into brigades to guard roads and catch spies. Sometimes these small militiamen carried spears with red tassels. They took their guard duty very seriously. It was comical to see a six-foot lad, armed with a pistol and carrying a dispatch case, pushing his bicycle vigorously through a city gate only to be stopped by a small boy, not quite three feet high, who would run up behind and ask, "Where are you going?" The big fellow would turn around abruptly, as if to say, "None of your business," but, catching sight of the determined little face below, would quickly smile, look resigned, glance at the bystanders as if to say, "What can I do?" and reach deep into his padded clothing for his pass. The boy would unfold it but since he couldn't read he would take it with dignity to some older person who would pass it around until somebody was found who could read it. If it sounded all right, he would wave the traveler on.

Everything that had been achieved in the Communist areas had been carried through by the united efforts and hard work of the people. The people's brains and muscles were the primary resource. By modern standards, China had not begun to industrialize, but compared to the Liberated Areas, Kuomintang China was an industrial giant with trucks, ships, trains, and planes, mills, mines, and power stations. Fast mail, telegraph, telephone, and radio linked the major ports and cities. Here a telephone line was a rarity and it was usually patched with small lengths of wire, many of them not more than two or three feet long. The main lines looked like the knotted wires we used to use in America to plant check-rowed corn. In spite of these handicaps, people were in touch with each other throughout the area. Carts and bicycles traveled the highways, newspapers were distributed, mail was delivered, and when something important happened every villager was informed by megaphone.

Newcomers to the Liberated Areas were often startled, as I was my first evening, by a noise which sounded like a combination of a dying cow and a donkey calling his wife. I found that this was the village announcer shouting through a megaphone from the top of the highest roof. I sometimes doubted the effectiveness of this means of communication, for whenever I heard the noise I used to ask those nearby what was being said. They invariably answered that they couldn't make it out. On one occasion, however, it did work, without a doubt. I went to a small village in search of a man named Han Ta-ming, labor hero number two of the Chin-Chi-Lu-Yu Border Region. The village mayor said Han lived a half a *li* away in another small settlement, whereupon he proceeded to the roof of his office with shouting tube in hand and summoned Han with a bull-like bellow. In a few minutes Han appeared.

This Han was typical of the men who had come to the fore during the years of war and revolution. He was forty-three years old. He stood six feet tall and was lean and strong. He talked freely of the trials he and his village had been through, and with pride of the way they had overcome all obstacles. Once he had owned ten *mou* of very poor land. Then the Japanese came, and after them the flood and the drought. The peasants sold their stock and tools in order to live a few months longer. Then they died by the hundreds. Han refused to die. He tried eating the seeds of weeds and grasses and found some that were nourishing. He organized groups to hunt for these. Those who survived went on to reclaim land under Han's leadership. They pulled plows with yokes around their necks. They collected tons of wild grass for sale as fuel in the city. They fought the Japanese. Then came V-J day, the Civil War, and land reform. Now Han had over thirty *mou* and a strong bull which he won when he was chosen as a labor hero by his neighbors, by his sub-district, by his county, by his sub-region, and, finally, by a conference of labor heroes of the whole Chin-Chi-Lu-Yu Border Region. Now, though he still lived and worked on his own small farm, he was a local leader and a national hero.

As I talked to Han, the other villagers gathered around eagerly and joined in the discussion. Each had something to add to what Han Ta-ming was saying. When the talk turned to the future, to the land they expected to reclaim, to the mutual aid they were even now organizing, to the coming winter production, to the possibility of a tractor or brigade for their sub-district, they all began to talk at once, as if a spark had suddenly caught fire.

Talking to these people one could not help feeling that the land had a future. After centuries of stagnation, they had suddenly started to move forward, perhaps not all at once, or at the same speed, but this year, life was better than last, next year was already on their minds, and they were thinking about a thousand new ways to increase production, to add to their land, and to work together.

This new mood was the result of the land reform and the new lease on life it had given the majority. The outstanding fact of life here was that the land and wealth derived from land had been divided. Land distribution was a reality and the people were for it. Those who had always worked the land now owned it. Ask anyone on the road and he would tell you: "Our family got a cart, and a three-section house," or "We got a cow, five *mou*, and eight bushels of grain," or "I got two *mou*. I took the rest in cash."

The people were behind land reform and they were willing to fight for their right to own a few *mou*. On a trip into the mountains, I made my way through 600 young men coming from the opposite direction. They had their bed rolls on their backs and a few utensils strung on their belts. They walked in groups of two and three, talking and laughing. They were the volunteers from but one sub-district on their way to their county town to join up. There was not one man in uniform among them, nor any armed escort to herd them on their way. The only weapon in the crowd was an old rabbit gun almost six feet long that one of the recruits was bringing with him as a personal contribution to the fire power of the army. A three-day campaign had brought 4,000 volunteers in that one county alone. Many more were turned down because they were needed at home, their health was poor, or their character unsavory. Only the very best men could get into the revolutionary army in 1947.

And that was something to think about in a land where armies had traditionally been built by tying ropes around defenseless peasants' necks.

The original plowing project in Chihsien closed down in the fall of 1947 when UNRRA was dismantled all over the world. Since gasoline could no longer be obtained to power the tractors, all the machinery was hauled away to the mountains and stored in loess caves. Students and staff dispersed to take up a variety of occupations, none connected with agriculture or farm machinery. I myself went to teach English at Northern University, a guerrilla institution of higher learning housed at that time in a small village called Kao Settlement in Lucheng County on the western slope of the Taihang range. While there I had the opportunity to join a land-reform work team as an observer and gather material for the book *Fanshen*.* In 1949 fuel for tractors became available and the plowing of wasteland started up once more. I was transfered from my teaching post back to the tractor project and settled down to the long hard task of China's reconstruction—that ten-thousand *li* Long March which Mao Tse-tung had warned would follow victory in the revolution.

The following pages tell the story of the first sixteen months of that Long March on one agricultural front—the training class at South Ridge, the establishment of Chiheng State Farm, the breaking of its wastelands with tractor-powered plows, the sowing of the first crop of wheat with modern drills, and the first mechanized harvest in North China. All of this took place on the eve of and immediately following that turning point in China's history, the smashing of the thirty-year-old rule of the Kuomintang and the establishment of the People's Republic of China.

* *Fanshen: A Documentary of Revolution in a Chinese Village* (New York and London: Monthly Review Press, 1967).

Chapter 1

Mule Cart to South Ridge

THE RUBBER-TIRED MULE CART that carried me and my belongings from Hengshui station bumped and swayed across the frozen ruts of the South Hopei highway like a dinghy in a choppy sea. Each time the two-wheeled vehicle hit a ridge or a hollow the heavy shafts pushed the rear mule sharply to one side. He had to shift his weight quickly and take the thrust of the cart on his shoulder or be thrown off balance. The lead mule, out in front of the shafts and hitched only by two stout ropes, strained forward unconcerned about the buffeting his mate was taking.

They were a handsome team, fat, sleek, and well groomed. The pride their driver took in them was evident from the red tassels on their halters and the shining bells that kept up a constant brittle jingling. The driver himself sat half sideways on the left shaft of the cart, feet dangling toward the road, hands crossed deep inside the sleeves of his padded jacket. His whip, tipped with a long lean thong that could be made to crack like a rifle shot, lay idle in the crook of his arm while he urged the mules on with a high pitched "Brrrrr-r-r, brrrrr-r-r-," rendered by rolling the tip of his tongue against the roof of his mouth.

Not that the animals needed any urging. Since the load was light and the air brisk they were in high spirits and kept up a rapid pace that broke, every so often, into a smart trot.

This cart and team spoke well for the tractor training center toward which we were heading. "Whoever is in charge evidently believes in doing things right," I thought to myself.

As we proceeded southward across the flat countryside made barren by winter I chatted with Hsueh Feng, a staff member from the center who had been sent out to meet me. He was a jovial

29

fellow, about thirty years old, who burst into song from time to time and made puns which my Chinese was too poor to understand. He wore the typical costume of the Border Region cadre—thick-soled cloth shoes, faded gray homespun pants and jacket, heavily padded with cotton, and a padded visor cap. With the ear-flaps of the cap tied under his chin and the bottoms of his trousers bound tightly to his ankles he looked, from a distance, like a skier at some northern New England resort. Closer inspection dispelled the illusion. Hsueh Feng's suit showed the results of many weeks of hard use. It had long since adapted itself to the contours of his body. Deep creases marked the elbows and the knees, and the cloth, since it could not be washed without taking the whole garment apart and removing the cotton, was darker wherever the wear was heavy. From the shine of grease on the sleeves and chest it was clear that he had been working with machinery.

"Are the tractors running already?" I asked.

"No," he said. "We towed them here with mules, all twenty of them. It was quite a job. We didn't even try to get them started. We have been waiting for you for that."

"I hope I can live up to expectations," I said, wondering what condition the machines had fallen into since I had left them in Chihsien in the fall of 1947.

Hsueh Feng spoke enthusiastically of the work ahead. When Tsinan, the capital of Shantung, had fallen a few months before, large stocks of gasoline had been captured. This brought the question of farm mechanization to the fore. Land reform was completed in all the old areas, and mutual aid groups were springing up on a mass scale. Soon the peasants would be calling for new tools and new ways of doing things. State-operated farms were going to be set up in the wastelands to pioneer new methods and apply machinery to agriculture. This was not to be relief or plowing for resettlement, such as UNRRA had initiated, but the inauguration of mechanized farming on a large scale as a commercial operation. Soon hundreds, perhaps thousands, of tractor drivers would be needed. And so, for a start, seventy students had been gathered together and gasoline was being hauled by mule cart from Tsinan. Only an instructor was needed.

"Now that you are here everything is fine," Hsueh said. "It was quite a job prying you out of that University. If it hadn't been for Chairman Yang, who intervened personally, I don't know if we'd have got you."

"But that's what I stayed on for in the first place," I said. "To do tractor work."

"I know, but try and tell the University that. Never mind, you are here now. Have you any plans for the course you are going to give?"

"No," I answered, taken aback. I had somehow expected to get down to that later.

"That doesn't matter. There'll be plenty of time for that," he assured me. "The main thing is to have somebody around who knows something about tractors. None of us in the Agriculture Department has ever seen one.

"The second day after I arrived they asked me to help make a plan for the gasoline, parts, etc., needed for the training school and the practice plowing that will go with it. Well, if I had studied agriculture or even engineering, I might have had some idea of how to go about it, but me . . . I was transfered there from the Drama Corps."

"How did they come to put you in the Agriculture Department?"

"I once worked in a factory. That was years ago in Honan after I ran away from my father's home. He was a big land-owner."

He laughed and I laughed too. In this New China you could never tell what might happen. There were so many projects opening up, so many thousands of jobs waiting for everyone. Hsueh's situation was far from unique. There were many people doing jobs they had never imagined doing and often doing them on a larger scale than anyone had ever dreamed of. The result was a vast surge of brains and creativity into all fields.

"What happened to the land we plowed here in '47?" I asked.

"Well, they got a crop of wheat, but it wasn't a very good one, and some of the land has already gone back to waste. There is a lot more land to the northeast that you never plowed. That's where the headquarters of the new farm is to be and that's where our training class is now—right on the edge of Chienchingwa, the Thousand Ching Basin.*

The sun was setting in a clear winter sky as we headed across the Thousand Ching Basin. Fifteen miles and four hours by cart from Hengshui station. Almost 17,000 acres of land, laid down by the disastrous floods of the early forties, lay as flat as a lake. Large stretches were overgrown with tough witch grass the peasants had been unable to control with their light plows and hand hoes.

As far as the eye could see everything was the same dull color—the yellow brown of North China earth. The leaves had long since fallen from the trees. Even the shoots of wheat, which might have shown a tinge of green on scattered patches, lay flattened and were covered with a coating of fine yellow dust. On every side the bare earth stretched to the horizon and the villages, whose walls and roofs were made of the same brown stuff as the plain itself, merged indistinguishably into it. The sky seemed enormous.

"What's to prevent new floods from ruining the land the tractors plow?" I asked.

"The dyke project," Hsueh answered, and his eyes lit up. "During the last two years a lot of work has been done on the river to hold it in its banks. The chances of a flood are much less than ever before. That's why the Ministry chose to start here. This time it is not a matter of a catch crop. They want to build a real farm here. In fact, there is already a section of land cultivated by the Regional Agricultural Bureau and Old Yang Pang-tzu [Big Yang] is farm manager. He is a former middle peasant from Nankung, a leader in the guerrilla war who studied two years in Yenan."

Far across the flat, against the sinking sun, we could just make out a high tower on the edge of the basin. "There it is," said Hsueh Feng. "That's South Ridge settlement. That's the tractor center. That building was erected by a landlord who made a fortune as a contractor in Tientsin. See how it towers over the landscape."

It certainly did seem strange, a building like that here in the

---

* A *ching* is 100 *mou*. One thousand *ching* equals 16,666 acres.

middle of nowhere, in the very heart of the plain, miles from the nearest highway and the nearest town. We were two hundred miles south of Peking and half way between the Yellow Sea and the Taihang mountains. Clustered around the tower, almost invisible in the gathering dusk, were the low brick compounds of the former landlord's relatives, and beyond them the mud huts of the peasants.

When we finally arrived it was almost dark. In the afterglow of the sunset we could see that the building was only a skeleton. The four-story structure had been stripped of its flooring and joists, and now towered above the village an empty shell, its gaping windows still ironically barred. The tower stood as a gutted monument to a China that was already gone, even if its scarred remains had not yet been cleared away.

Hsueh Feng pounded on a large wooden gate. As it swung open on a creaking hinge, we saw oil lamps lighting up the courtyard inside. A crowd of young people swept out into the street, picked up our belongings, and surged with us back through the opening. Eager hands shook mine. Weather-stained faces broke into laughter.

"Han Chiaoshih [Teacher Han] has come, Han Chiaoshih has come." Soon the whole area was filled with people, all laughing and talking at once.

A little man in an outsize padded suit took me by the arm.

"How are you?" he asked in strongly accented English, and everybody laughed with delight at the effort. Then he introduced himself.

"I am Li Chih, director of the training school. We welcome your coming. We have been waiting impatiently for you for a long time. Welcome."

"Welcome," said the others. "Welcome to South Ridge."

Chapter 2

"I've Never Seen a Freight Train"

EVEN IN BROAD DAYLIGHT South Ridge lay lost and desolate in the vast expanse of earth and sky that stretched out to the horizon on every side. This no one seemed to mind, however.

Long before sunrise the population was up and stirring. Those who were not awakened by the crowing of the village cocks were roused by the clear notes of Wang's bugle. Wang, a student just out of the army, had a powerful pair of lungs and a keen sense of time.

The notes of the bugle had hardly died away when the cold stillness was again shattered by the full-throated singing of the students, followed by the steady one-two-three-four of the morning exercises, which they did in unison. This sound mingled with the cracking of the peasants' mule whips and the rumble of their iron-tired carts as they rocked down the rutted lane toward the basin. From the distance came the laughter of village women as they coaxed their blinking donkeys out of their warm stalls and set them to work grinding corn at the stone mills beside the street. In the half light, men and boys were already at work on the flat, scraping dirt into mounds which would later be processed for salt.

By the time I had risen and dressed, the students had already tidied their quarters, swept the courtyards clean, washed their hands and faces, brushed their teeth and hair, and settled down in small groups to a quiet half hour of newspaper reading.

At breakfast, which was eaten in an open courtyard around a huge cauldron of steaming millet, meetings, classes, and work details were announced, and before I had time to finish one bowl, everyone had disappeared to begin the day's work.

34

As I rushed to get to class on time I met the village children, slates in hand, running gaily if somewhat noisily off to the little adobe structure on the edge of the embankment that served as a school. The children were fortunate to have a building to meet in. There was no roof in South Ridge large enough to cover the tractor classes. A crumbling wagon-shed to break the northwest wind, a blackboard propped on an old cart, a tractor rolled up beside it, some logs laid on the ground for seats—that was the lecture hall of the Tractor Training Class set up by the Ministry of Agriculture and Forestry of the North China People's Government in the winter of the year 1949. Students who couldn't find a place on the logs brought bricks with them and sat huddled in groups wherever the wind blew less fiercely. On days when the sun shone brightly and the air was still the outdoor classroom was a pleasant place, but on other days it took all the fortitude the students possessed to stay there. The wind, blowing down on us right out of the Mongolian steppe, swept over the low roof of the shed and set the straw and chaff in the yard whirling in the air. The dust blew into the students' eyes and settled in a thin yellow film on the pages of their notebooks. It was so cold that the ink froze in the inkwells and fountain pens. The students' hands got so numb that they could not write, and every few minutes they had to stop to blow on their fingers. Some wore a mitten on one hand and wrote with the other.

Every half hour, Ding Wen-ch'ing, the class leader, blew on his whistle and they all stood up, jumped up and down to bring the feeling back into their numb toes, and beat their bodies with their arms to get the circulation going. The cooks brought a kettle of boiling water. The students sipped it as hot as they could stand it, and then returned to their tasks while the warmth spread momentarily outward toward their limbs.

At such times I used to watch their faces for any sign of discouragement. I never saw any, nor did I ever hear any grumbling. When the wind blew, the young men and women thawed out their pens, brushed the dirt from their paper, and went on taking notes as if nothing had happened.

Of course, not all of them could take notes. At least a quarter of the class was illiterate, or at least not literate enough to write

down the many characters needed to transcribe a lecture. Some
of the more advanced complained. They asked for a division of
the class, with courses matched to various educational levels. But
Director Li would not hear of it. He was pleased with the variety
of the students.

"Knowledge is not all contained between the covers of books,"
he said. "Some who can't read nevertheless know a lot about
machines, and others have raised crops. They can all help each
other. That's the way it should be."

Sun Hsiao-kung, a student from Peking who had come across
the lines into the Liberated Areas only a few weeks before, acted
as interpreter. I explained each concept carefully to Sun in
English. Once he understood the English, he then tried to understand what I was talking about. Only then was he able to explain
it to the students in Chinese. This task was complicated by the
fact that Sun had studied social, not mechanical, science, and
everything from a wrench to a cotter pin was new to him. He
knew neither the English nor the Chinese terms for the parts or
their functions. The few Chinese terms that he did know turned
out to be colloquial.

We soon discovered that there was no common mechanical
language in China. The names for parts and tools were as varied
as the regions where men drove trucks. What they called a piston
depended on whom they had learned from. The Northeasterners
(Manchurians) had learned their trucking from the Japanese, who
in turn had learned from the Americans. For them the parts had
meaningless English names like "pea-stone" for piston, "shi lin da"
for cylinder, and "ku lan ke sha fa ta" for crankshaft.

This would have made things very easy for me, once I got the
hang of it, if mechanics from other regions had had the same
system. But they didn't. The North Chinese, being more original,
made up names of their own, most of them very apt and descrip-
tive. We could have used these except for the fact that the
Chinese in the Yangtze Valley, where trucking and the machine
industry was more highly developed than elsewhere, used a different set of names.

What we finally called the parts depended on what the "technical group" decided. The technical group was made up of five

or six students who had formerly driven trucks, worked in repair shops, or handled machinery of some kind. If they couldn't agree on what to call something, Sun chose the word that sounded best to him. So piston ended up as *kou pei* (plunger), cylinder as *ch'i kang* (gas chamber), and crankshaft as *wan chou* (bent shaft).

Fortunately, the course did not depend on lectures for its substance. The core of our activity centered on tearing down tractors part by part to learn their names and functions. A lot more time was spent unbolting things, removing head and pan, and watching the pistons move up and down, than in talking in front of the blackboard. It was not long before Sun became an enthusiastic mechanic and lectured brilliantly to the class on things he had learned an hour or two before.

The technical group played an important role. They helped prepare material, put on demonstrations, and took turns teaching the things they knew. Had I not been there, there would still have been a course based on the collective experience of these workers.

The mechanical side of things came easily to the students: they could see how the parts fit together and how they moved. But to explain something like electricity was a different matter. I never was able to make it clear, perhaps because it had never been really clear to me. We spoke of electricity by analogy, but none of the analogies could be extended too far.

One day in class one of the technical group asked me, "Which end of the battery does the spark flow out of?"

"It doesn't flow from either end," I said, "It moves simultaneously through the whole circuit at once."

"But there is a spark wire and a ground wire," he said.

"That's just for convenience sake. We call part of the circuit the ground and part of it alive. It could just as easily be the other way around."

"But when we hold the spark wire near the block we can see the current jump," he said.

"Which way does it jump?"

"From the wire to the block."

"You can see that?"

"Sure."

"But look," I said. "That's impossible. The spark is instantaneous. It exists at both ends and the middle at the same time."

"But you said it flowed."

"It does and it doesn't. It is not like water flowing through a pipe. It's like the shock that goes through a train when it starts up. Each car hits the next—bang, bang, bang, and the energy passes down the line."

The class looked puzzled.

"Haven't you noticed how a freight train starts up?" I asked.

There was an awkward silence. I overheard a young woman turn to the student on the log next to her and say: "I've never seen a freight train."

The idea behind an induction coil, the concept of magnetic lines of force cutting across wires, or of wires cutting across magnetic lines of force, was even harder for them to grasp.

Since we had no magnet I took an iron rod, wrapped it with wire, and connected the ends to the poles of a battery. To the delight of the class this rod picked up lumps of iron and made iron filings bristle on a piece of paper. But before everyone had had a chance to see and understand what was happening the wire got so hot it burned my hands. I dropped the bar. When we hooked it up again the battery, which would have to be taken all the way to Shihchiachuang by train for recharging, had lost its juice.

Partial understanding never satisfied the students. They wanted to know every last detail. They wanted to take everything apart, even things that could not easily be put together again, like the coil. It was not enough to give them a general idea of how a carburetor functioned. They wanted to know what each hole was for, why this one was larger than that, exactly where the air went in when the engine was idling and where it went in when it was racing.

"What is this hole for?" they asked.

"That is the air intake."

"But the little one?"

"That is also for air."

"Why is it so small?"

That had me stumped.

"The engineers have worked it out experimentally, they find that that size works best," I said, trying to make the best of it.

"But that's no answer," said the students. "Why does it work best? What would happen if it were larger?"

"The truth of the matter is, I don't know," I said and we all laughed.

Such questions were saved for the technical group. We used to meet until after midnight trying to find a satisfactory explanation to all the questions. Bit by bit, without instruction books or texts of any kind, we worked out the proper function of every hole and passage in the carburetor, every wire and connection in the electrical system, and every spring and lever in the hydraulic pump that lifted the plow. Everything was studied and argued out.

The hydraulic pump gave us no end of trouble. We took it apart, studied it for several days, and learned everything except what made it stop when the plow reached the top. For that there were many different theories, not one of which made sense. It was only many days later that one of the students, working on a jammed pump in the field, found the answer.

I would have been ashamed of my ignorance had it not been a stimulant to searching study on the part of the students. Instead of telling them everything, I had to lead them to discover the tractor's secrets. Thus the lessons were really learned instead of memorized and forgotten.

In the old society it would have been considered a terrible loss of face for the teacher not to know something. But to Director Li and the students "face" was a stupid hangover from the past. One of the things they liked about me was that I frankly told them when I was stumped.

"Han Chiaoshih does not care about 'face,'" they said. "When he doesn't know something he says so."

The common search for knowledge put us all in the same boat and made our work collective. I was not *kao kao tsai shang* (high high on top) but down on the ground with the rest.

Collective study was taken for granted. It had a long tradition in the Liberated Areas and was rooted in the common need of the people to find ways out of a desperate situation. During the terror launched by the Japanese and continued by the dispossessed landlords' "Home Return Corps," prima donnas and individual heroes were of little use. Nothing less than the combined brains and energies of the whole community could protect them. The tunnels dug by the peasants of Central Hopei that connected the villages underground and made resistance on the flat land possible, could never have been built without the labor and ingenuity of millions working together. And so there had grown up the habit of consultation and the pooling of all experience, ideas, and energy.

In his weekly talks to the tractor students, Director Li made this very concrete.

"What is the use of training a few smart drivers if the rest of us lag behind?" he asked. "Can three or four operate a ten-thousand-acre farm? What is the use of one man taking good care of his machine if the next man to go on shift hasn't learned his lesson well and ruins it? We must develop a team. Everyone must be helped to become competent."

Those who could take notes helped those who could not. Those who could draw made diagrams for those who found it difficult. Those who had some experience with machines worked patiently with those who had none.

In the evening I used to go from compound to compound to help the students with their study. They lived in several dilapidated courtyards that had once been occupied by a landlord's in-laws. The dirt floors had been swept clean, new paper pasted on the window lattice, and a thick mat of straw laid down for bedding. In some of the rooms there were old brick sleeping platforms that could be heated in winter, but for the most part the students had to sleep on the floor. They were packed in ten or twelve to a room. Each person's quilt was rolled up against the wall. Towels, toothbrushes, and extra clothing hung on nails. Few had more than the clothes on their backs to worry about, so the rooms were not cluttered.

After dark I always found the students clustered around an oil lamp in groups of two, three, or four, heads together, notebooks on their knees, vigorously discussing. They read aloud from their notes and tested each other. "What are the four cycles?" "Why must the spark be advanced?" "Why does the piston have three rings?" And each time someone opened his mouth his breath condensed in the cold air.

Often groups came to me for permission to borrow carburetors, distributors, or oil pumps from the parts department. They brought the parts back to their quarters and took turns explaining them to each other. "Now you see," said one, "this is the carburetor, this is what mixes the gas with the air."

"What do you want to mix air with it for?" asked another.

"Without air it won't explode, but it has to be just the right amount. The air comes in here . . ."

"What makes it come in? Why should it want to come in at all?"

"Because the piston goes down and that makes a vacuum. It is sucked in . . ."

And so they continued until the bugle sounded for sleep.

Chapter 3

Wasteland Pioneers

THE STUDENTS came from every class and background and from all over North China.

Smallest and most striking of the three women in the school was a mountain girl, Chi Feng-ying. Not yet twenty, she had a lovely oval face and a wide-eyed look of wonder. She wore a peasant jacket with cloth fastenings in place of buttons, and a white towel on her head after the manner of her home village.

When her parents had died of starvation, she had been adopted by a family only slightly better off who reared her as the intended bride of their son. She had suffered from hunger and exhaustion during the burn-all, loot-all, kill-all campaigns of the Japanese in Wutai and was subject to fainting spells and violent headaches. She wrote slowly and laboriously and persisted in studying long after the others had gone to sleep. Everyone worried lest her health give way completely, and she most of all, for she had set her heart on becoming a tractor expert.

The most efficient and capable of the girls was Li Chen-jung, a middle peasant's daughter. She wore her hair straight and pulled her cadres' visor cap down low over her forehead. Her nose was rather large and sharp for a Chinese and her mouth thin and short. Strongly built and full of energy, she naturally took the lead in everything, helped organize a branch of the Youth Corps among the students, and kept exemplary notes from which the others in her study group gained much help. She had been to high school and had been trained as a teacher, so she read and wrote fluently. Her shortcoming was a certain lack of warmth and a certain mechanical approach to problems. She always wanted to know how many bolts there were on the pan and how many on the head. She kept a careful record of these facts, though as far as I could see it served no purpose.

Ho Chi-ying, a Chengting merchant's child, was a lively, healthy, fun-loving girl, strong, broad of face, and solid of limb. She too had studied to be a teacher. She knew her characters and her figures and was quick to catch on to anything new. She was the best of the three at working with her hands.

Among the boys there were all sorts: intellectuals from Peking universities, workers and craftsmen from Shihchiachuang, the biggest city in the Liberated Areas at that time, peasants both rich and poor, landlords' sons, demobilized soldiers, and young government workers. About half of them were chosen through publicly announced examinations. The other half were sent by various regional agricultural departments, army units, schools, and local governments.

A natural leader among them was Kuo Hu-hsien, former land-less peasant from the Taihang mountains. At the age of fifteen he led the militia of his village against the Japanese and killed more than a dozen of the enemy using homemade stone mines. Frail and soft spoken, Kuo weighed only a little over 100 pounds. Not yet twenty-two, he had the maturity and restraint of a man of forty. People sought him out and trusted him because he looked at problems dispassionately. He seemed to have no personal vanity, make no personal calculations, harbor no personal feelings to intrude into and distort an objective judgment. He had completely identified his own life with the cause of New China and asked nothing more than to serve where he was needed.

Quite different, and yet also a leader, was Chang Ming, a proud and brilliant young schoolteacher from a Wutai mountain town. As a son of well-to-do parents he had had many advantages and a good education. He was passionately interested in science, always wanted to go deeper into things than anyone else, and dreamed of being an engineer—a great engineer. That was the difference between Chang and Kuo. Chang had personal ambitions that sometimes created a conflict. He was impatient with those who were not as quick and bright as himself.

Ding Wen-ch'ing, elected leader of the student brigade, was a peasant and demobilized soldier. His character was as clear and open as the sky and his heart full of the joy of life. He was not given to subtle thought, but took his duties as brigade leader seriously and worked hard at anything he was asked to do. Like Kuo, he was deeply devoted to the cause of New China, but he did not understand as well what that meant. His outlook on life was simple, yet he had a peasant's shrewdness where material values were concerned. No one ever got the better of him in a commercial bargain, and that was a good thing for the tractor work, for many deals with merchants had to be negotiated.

The class also included Liu Kai-chang, who stood six feet four in his cloth shoes but insisted that in his home village he was short; Lao Pao, who was always angry at something; Wei, a former medical orderly in the Eighth Route Army, who made everyone laugh at his assorted mimicry, as he applied first aid to cuts and bruises; Li, an intellectual who was always vaguely dissatisfied; and Ma Lien-hsiang, who sacrificed his chance to study in order to take over the direction of the kitchen so that everyone could eat as well as the meager budget allowed.

Taken together they reflected a fair cross-section of North China youth, neither better nor worse than the average, though perhaps, on the whole, a little more adventurous. They came to study tractors for a variety of reasons: some because they had been sent by their home organizations, others because they were bored with what they were doing, still others because they believed wholeheartedly in the power of machines to transform the society in which they had grown up.

T'ang Yu-ming came because he thought tractoring would lead to a truck driving job and hence a chance to travel and see city life. Chang Ming came because he saw it as an opening to an engineering career. Kuo Hu-hsien was sent by the Agricultural Department of the Taihang Mountain Region because he had a vision of what mechanization could do in the mountains. Li, the intellectual, came because the whole setup sounded romantic.

There was no doubt that romance surrounded the very idea of tractoring. Tractors had become a symbol of all that was new and bright in the countryside. The "iron ox" would draw in its wake a whole new world. A shimmering aura of prestige and progress enveloped mechanized farming and drew young people as a magnet draws filings. In China there was no occupation, except flying, which carried with it such public interest.

Those who came to learn found out, of course, that tractoring was not all romance, that in fact it consisted mainly of hard, monotonous, greasy work. Nevertheless, none of them quit. Of the seventy who started out that year, not one left for home or asked for a transfer to less arduous work. And this was due, I think, to the common background shared by them all.

What they had in common was the experience of the last decade, the experience of China's war of resistance and the land revolution that followed it. They had all been through an invasion as ruthless and terrible as any country had ever suffered, and they had seen the peace, so bitterly won after eight long years, smashed by another invasion. This second invasion, into the Liberated Areas they had all helped to build, was attempted by China's traditional gentry, led by Chiang Kai-shek and armed and financed by the United States. The students knew that their country stood at a crossroads in history, that smashing this invasion could end, once and for all, the misery of feudalism and foreign domination. If the invasion succeeded they would return to the past.

Not one of them held any brief for the past. In their minds it stood for greed, graft, and corruption; debt, rent, and backbreaking toil; buy-and-sell marriages, concubinage, and infanticide; national humiliation and betrayal. Those who came of poor peasant stock hated the past. Even those who, as landords' sons had benefited under the old system, were sick of it because of the humiliation it had brought on China. They passionately desired something new.

It was Director Li's task to deepen this passion and to direct it into a disciplined, self-conscious effort to build a new type of agriculture. If the program was to be a success, he had to weld this multiple assortment of individuals into a strong collective force, into a battle-seasoned team capable of conquering all natural obstacles and of resisting all the corrupt influence of the old society which still surrounded them and from which they had all sprung.

"Our task," Director Li told the students, "is to build islands of socialism in a vast sea of individual farming. We are the ones who will have to show the way to the whole country. It will be no easy job and it will not be accomplished in one year, or two years, or ten years. Chairman Mao has told us that nationwide victory for our armies is but the first step in a ten-thousand-*li* Long March to build a new China. We are pioneers on that march."

Chapter 4

Director Li

IT SOON BECAME CLEAR that Director Li was the dynamo around whom life at the school centered.

In appearance he was not particularly impressive. A Szechuanese of medium height, lean, dark, and quick in his movements, he contrasted rather sharply with the more stolid, muscular peasants of the North China plain, who made up the majority of the student body. Director Li's clothes, especially his padded winter garments, were too big for him. He seemed to walk about inside his enormous pants. The cuffs of those pants had to be rolled up to keep from dragging on the ground and the baggy seat always seemed to float halfway to his knees. On his feet he usually wore the half-leather, half-canvas shoes peculiar to the Liberation Army of those days, and these shoes, like the rest of his clothes, appeared to be too big. Since both left and right were made from the same pattern, they lent a rather comic touch to the whole outfit. There was much about Director Li's appearance that reminded one of Charlie Chaplin.

Out of the jacket, which padded out Li's shoulders and hid his slender build, protruded a rather scrawny neck attached to an alert head, made to seem small by a bulky padded cap. From under the cap a shock of long black hair stuck out. When this fell over his face he removed the cap, flung the offending lock back over his head, and smoothed it into place with his right hand.

Director Li had a warm friendly smile that revealed very prominent upper teeth and a slight twist to the muscles of his face. This caused one corner of his mouth to rise higher than the other. But the two things that one noticed more than anything

else were his large, bright, active eyes, and his graceful, slender
hands. Li used his hands as he talked and the tendons stood out
sharply on their backs. His eyes revealed an intelligence that was
to impress me deeply as time went on, while his hands suggested
a manual skill that few Chinese intellectuals possessed.

Director Li was one of those people, rare in old China but
springing up in increasing numbers in the new: an intellectual
with his feet on the ground. The anti-Japanese war and the revolution against feudal control and domination brought the best of
China's educated young men and women into close contact with
the people, where, faced with problems of life and death—famine,
guerrilla warfare, the scorched-earth campaigns of the Japanese
and the murderous rampaging of the Home Return Corps of the
landlords—they learned to adapt their academic training to the
rock bottom activity of organizing peasants to fight for their
very existence.

Most of these young people eventually became Communists.
They gave up their homes, their land, their wealth, sometimes
even their fiancés or wives, and went into the mountains to devote
their lives—to sacrifice them, if need be—for a New China. In
doing this they found a life immeasurably richer, finer, and
warmer in human relationships than any they had left behind.
Their new existence was based on companionship and unselfish
work together with millions of ordinary people. In the course
of this work they themselves were transformed until they retained
hardly any trace of the selfishness, vanity, frustration, and sophisticated ennui that had characterized the lives of so many educated
Chinese in the past.

Director Li's family was a bankrupt branch of one of the
wealthy Szechuan salt-producing dynasties. He was born and
grew up among the salt fields where for centuries deep wells
dug with ingenious bamboo bows had tapped pools of salt water
many thousands of feet under the earth. Long before the West
invented deep drilling, the Chinese had developed techniques for
cutting through rock to almost any depth desired. The Li family
had been in the business for generations.

Li was a precocious but rebellious youth. When he reached

high school he was expelled from one center after another for political activity among the students. At that time he did not have any fixed views: he only knew that something was basically wrong with China. For a time he and a group of friends became convinced that cooperative farm communities would save the peasants, so they all transferred to an agricultural school to study bees, vegetable gardening, sheep, and dairy cattle. After a year Li decided that without revolution agricultural cooperation was nothing but a Utopian dream and transferred somewhere else. However, he never forgot the smattering of technical knowledge he had picked up during that period.

"It comes in very handy now," he said to me. "You see, all those different things I did in my youth—they have not been wasted. Sooner or later the time comes when what you know is needed."

Very early he became interested in mechanical devices. When bicycles made their appearance in that out of the way region he was the first to own one. After he learned to ride he taught the hired help around the school, the floor sweepers, water carriers, cooks, and their apprentices. In time he built up a big following among them.

"I told them never mind if anything breaks, just come and tell me and we'll take it apart and fix it ourselves and we'll learn just that much more about it. So we all became skilled bicycle repair men and skilled riders too. I was never stingy with that bike, I didn't keep it to myself, and the result was that the others were very responsible with it. They took very good care of it.

"As it happened, that bicycle group turned out to be very useful to the student movement. Just at that time the students in Szechuan were very active. The Japanese were moving into North China after taking over the whole of Manchuria, and the Kuomintang was reacting by arresting those Chinese who protested. We students got very angry and staged a number of mass strikes to force the government into taking a stand in defense of the nation.

"Well, of course the strikes were suppressed. The governor sent troops to guard the gates of the schools and universities and tried to prevent the students from contacting each other. But we managed to keep in touch just the same. A lot of people wondered how we did it.

"The secret was my bicycle group. No one suspected that those boys had any connection with us. They came and went at will and so we were able to send messages anywhere we wanted to."

Director Li went to almost as many colleges as he did high schools and gradually his political views became clearer. He began to see that it was the feudal land system itself and the foreign oppressors who propped up that system that made life so terrible for the people. He began to read books on Marxism and eventually joined the Communist Party in order to work for the liberation of his country and the end of feudal backwardness.

Then began the years of hard, dangerous organizing work among the people of his native province. He worked for a time among the salt workers of his home county. Then he was employed as a clerk in the postal service truck repair depot in Chungking. Still later he went into the mountains of western Szechuan to work among the Tibetan mountaineers. While there he worked in the branch office of a bank. When things got hot for him, when the authorities were about to investigate who was behind the progressive activity in the area, he moved on, only to start all over again in some other place. In spite of this uncertain life, he found time to marry a young student, Hsiao (little) Fan.

"Several things helped me a lot," he said. "One was my name, a very distinguished one around home. The other was my position in the Ku Lao Hui. That was the strongest secret society in West China. My uncle had been a leading man in it and I inherited his title. Of course, I had to be clever enough to hold on to it. But a position like that meant great power. I commanded the services of hundreds of men. The society was really a front for carrying on the intrigues of the reactionary landlords, but it had revolutionary origins in the revolt against the Manchu conquest and a lot of its members were poor workmen, gang-ridden stevedores, carters, and boatmen from the river ports. The organization was especially strong among the boatmen and since the rivers are the main arteries of our province that meant a lot to me.

"We had signs and signals to show who was a member and

what rank he held. All I had to do was walk into a waterfront eating stand, put my chopsticks down in a certain way, and I was immediately in touch with men who could get me a boat to any place I wanted to go."

In 1945, after the Japanese surrender, the Kuomintang and the Communists held talks in Chungking. To reinforce Chou En-lai's Chungking staff, a number of Szechuan province Communist Party members were called from the underground. Li Chih, who by that time was already the father of a daughter, was one of these.

When the talks failed, the underground members could not go back where they had come from. They would have been arrested, perhaps shot. Chiang Kai-shek did, however, guarantee them safe conduct through the lines to the Yenan Border Region. They traveled with their families north to Yenan by truck, never sure whether the notorious General Hu Tsung-nan, who commanded the blockade of the Yenan region, would honor the safe conduct or not.

Director Li stayed on near Yenan until Hu Tsung-nan invaded in the spring of 1947. Then he and his family started on the long trek through North Shensi that lasted until the 300,000-man Kuomintang army of the Northwest was finally routed and destroyed twelve months later. The Li family baby girl traveled in a basket hung on one side of a donkey, while all the Li family household possessions balanced her in a similar basket hung on the other side. The travelers stopped at countless out of the way mountain hamlets, worked for a while, and then moved on, never more than a few miles ahead of the invading troops.

Near the city of Suiteh, Li worked for a while as finance officer for a mountain county. He left through the north gate with the county funds and records packed on mules when enemy patrols entered the town from the south. The next day he received an order to join a guerrilla detachment that was harrassing the Kuomintang regulars as they advanced. He had to leave his wife and child behind. Hsiao Fan was pregnant again and due to give birth within a few days. Since the area was almost sure to be occupied by the enemy, the only thing to do was to disguise her in peasant clothes and leave her with some local family in hopes that no one would discover that she was a cadre's wife from far away.

With a heavy heart Li rode off to his first battle.

"I hated to leave them behind," Director Li said, "but I thought the best way to protect her was to put up an active fight. Then the enemy soldiers wouldn't have time to go snooping in all the caves.

"I didn't know what I would do when I got to the unit. I had never handled a gun before in my life. Our fifteen men had no sooner settled on a big knoll overlooking a bend in the road than we saw enemy patrols advancing. They peered right and left as if they expected every stone to explode. There were at least one hundred of them. We let them pass and sent word back to the regulars. A few hours later when our own troops attacked we went in to the enemy's flank and harrassed them as they ran. That was the beginning of Hu Tsung-nan's big retreat. In a few days the whole area was cleaned out and our army went on to liberate the whole Northwest.

"On the night of the big battle I got a message from my wife. The enemy troops had gone right past the door but no one had searched the house. She was safe and we had another daughter.

"That was the beginning and end of my military career. Hsiao Fan has never forgiven me for going off that day and leaving her at the mercy of the enemy. But there was a principle involved. Where would we be if everyone looked first after his private affairs and tried to play safe alone? The only way is to stick together, work together, and act in a disciplined way. Only because we all went out and fought were we able to drive those bandits back and bring safety to everyone."

From North Shensi, Li and his family came east to join the land-reform work in the Chin-Cha-Chi region west of Peking. This time the donkey carried a baby in each basket and the household goods had to be piled between. Li served as secretary to a land-reform team that was in charge of a whole county. When that job was finished he went to Shihchiachuang to await a new assignment. While there he read in the newspapers about the tractor school. He applied and was accepted.

"But," he said, "I applied to come here as a student. Instead they made me the director. What a break. I wanted to join the classes and learn all about machinery. And now I haven't time."

Whenever he had a spare minute, however, Director Li came to class and sometimes in the evenings when we both had nothing to do he asked me to explain to him the parts that he had missed. He had a flair for mechanical things and, like the students, was keenly interested in every detail. He remembered a great deal about the postal service repair shop he had worked in in Szechuan years before and was always comparing notes.

When it came time to put the tractors in working order I found that we had no distilled water to use in the batteries. As there were no stills in the vicinity, I thought we would have to send to Shihchiachuang for one. Director Li would not hear of it. "I'll make one myself," he said. He spent one evening with paper and pencil sketching out plans. The next day he walked into town and came back a few hours later with a little still set up and ready to go.

"Where did you get that?" I asked, both surprised and happy.

"The local tinsmith and I put our heads together and invented it," he said. He was very pleased with himself. He called the students together and showed them how the still was made so that they could make one themselves in the future.

That was typical of Director Li. To him the whole of life was education. There were lessons to be learned from almost everything that happened if only one took advantage of it.

I have never met anybody who enjoyed young people more than Director Li or who helped to develop in them more zest for life, fun, work, and study. One reason was that he himself enjoyed living so intensely. Every task was for him an adventure to be savored to the full and shared with everyone. Whether it was designing a still, planning a course, analyzing current affairs, or helping the cook to make bettter meals, Director Li's own enthusiasm was so infectious that he carried everyone else along with him.

He loved to be in the thick of whatever was going on. Whenever possible he took part in the late afternoon basketball games that were played on the threshing floor behind the tractor shed. In spite of his small size he proved to be no slight opponent.

Quick, clever, and tireless, he was always waiting where nobody expected him. Between exhausting meetings he used to relax with a hand of poker. This gave him a chance to meet and talk with students in circumstances that had nothing to do with work. When dancing came to the countryside he was one of the first to learn. As with most of the young people, it was a matter of indifference to him whether he danced with a man or a woman. What he enjoyed was the music, the crowd, and the social atmosphere.

It did not take him long to promote an orchestra of Chinese instruments, an opera group, and a drama corps at South Ridge. He believed in hard work but he did not believe that life should be all labor and no fun. This endeared him to young people immediately.

Director Li was also a good listener. He believed that everyone had something to contribute, something valuable to say. When problems came up for discussion, he was always the last to speak. He presented the issues, then urged everyone to talk them over. Sometimes the group came to a conclusion without his saying a word, If it was a good one he accepted it and put it into practice. If the discussion got entangled and came to an impasse he always came up with a fresh approach, a new insight that cut through the difficulty and enabled the group to go on and solve the problem. He was able to do this because he concentrated on the matter, listened intently while others were talking, and studied what they said with an open mind.

Another thing that impressed me about Director Li was his complete lack of concern for himself. He was not a healthy man. He had piles and a chronic liver infection, but he never asked for any privileges, any special treatment, any special food. He lived off the same subsistence allowance as everyone else—food, clothes, and fifty cents a month spending money—and worried only that others were not getting enough. He was always asking the cook to make a plate of eggs for me. "Lao Han is not used to our food," he would say. "And besides, he has a big body and needs extra nourishment." But he would never eat any of the eggs himself.

When I arrived at South Ridge he offered me my choice of

rooms. I chose the west side of the farm courtyard. This was the room with the biggest window, the most sunlight, and the dryest floor. Most of the dwellings were low and damp, with only the tamped earth of the plain underfoot. But the west room was built about three feet above the ground and paved with large square bricks. It turned out that he himself was living there. He moved out without the slightest sign of annoyance when he might easily have said: "That's mine."

During the winter of 1949 Li's wife was at the seashore sick, with another baby due. By that time he had four daughters, all of whom were in nursery schools run for the children of government cadres. He did not regard the separation from his wife and family as an imposition. There was work for him to do in South Ridge. "The government takes excellent care of my little girls," he said. "I don't think I should be more concerned for their welfare than for the welfare of other people's children. We are working for all children so that they can grow up in a new world that will belong to all of them."

Chapter 5

Lao Hei Gets Married

ONE OF THE STUDENTS who could not read or write his name but nevertheless had a lot to teach the others, was Lao Hei, a mechanic from Shihchiachuang. Handsome, strong, and straightforward, he was always breaking things and getting into trouble because of his enthusiasm. Lao Hei usually said exactly what came into his mind. This made the others laugh at him, but he didn't mind. He laughed right along with them. His chief worry was his inability to read. The characters he studied one day he forgot the next. This disturbed his natural self-confidence and brought creases to his broad forehead. Another problem that perplexed him was marriage. When he arrived at the school he was already twenty and had no prospect of a wife.

Lao Hei soon made a name for himself. Wherever anything had to be taken apart or put together, blocked up or knocked down, he was always on hand to do it. He had an athletic build and great physical strength. Having made such slow progress in reading and writing he was happy to be able to excel in muscular feats. His grin was always broadest when he had just moved something that everyone else said was too heavy for anybody to lift. "Call Lao Hei, Lao Hei can do it," became a common saying around South Ridge.

Lao Hei became very attached to me. I helped him with technical problems and he helped me prepare class material. One day as we dismantled one of the tractors, I asked him how he had happened to become a mechanic. He seemed much closer in temperament and habits to a typical peasant than to any of the mechanics I had met before. As a group they tended to be sophisticated and opportunistic.

"My father died when I was very young," he told me. "My mother scraped along for a few years, but when things got tough there was nothing to eat in the house. I was ten years old, but I was big for my age, so, with the help of a family friend I apprenticed myself to a Japanese truck driver."

"Did he teach you much?"

"He taught me as little as he could. He kept me to do the dirty work. When he said, 'Get me the three-quarter wrench,' I had to have it there by the time his hand reached out. If I didn't, I was beaten. He lost his temper fifty times a day. 'Get me two buckets of water.' 'Where's the oil?' 'Wash those rags.' 'Why haven't you wiped that fender off?' I was on the run from sunrise to sundown, and often far into the night. I had to get his food for him, and wash up afterward. And for everything I did wrong I was beaten with whatever he happened to have handy. Sometimes it was a wrench, sometimes a stick; sometimes he used his leather belt. See, here, the scar on my head—that's where he hit me with a wrench. When he got really angry he made me stand outside in the cold with only a cotton shirt on.

"I learned quite a bit. Couldn't help it, because he was lazy. Shoved more and more of the work onto me. But he was smart. Didn't show me the tricky things. He kept them to himself. After all, why should he teach me everything in one day? Even after three years I hadn't learned to scrape a bearing, or time the spark. When there was something like that to do he always sent me away on some errand."

"Did everyone treat apprentices that way?"

"Sure, that was the ordinary thing. They made you sweat for knowledge. They didn't teach you for nothing, and they never taught all they knew. Lots of mechanics had to take work back to their old masters for the final touches, and pay well to have it done, too.

"These students are lucky. You're teaching them everything— one, two, three. Why, it took me three years to learn what we went through this month. They don't know how lucky they are. The new society wants everyone to learn as much and as quickly as possible. That is something we never even dreamed of. And to think, I got all those beatings for nothing."

"But now you can help the others. You have so much experience, you can save them a lot of trouble," I said.

"Yes," said Lao Hei. "That's the way it should be, but they shouldn't take it for granted. They shouldn't forget how it was for us."

From Lao Hei and others I learned something about the truckmen's lives in the old days. In that world it was each man for himself. Each man's main concern was to protect his rice bowl and that meant to protect the secrets of his trade. If driving and repairing ever became common knowldege there would be a lot more competition and it would be that much harder to make a living. The idea was to restrict the craft and to build up the craftsmen into an elite who ruled the transportation world.

Opportunities were not lacking. In old China transportation was at a premium. It was easy to collect money from "yellow fish" (illegal passengers), steal gasoline, swap new parts for old, do favors for old friends or for the highest bidder, smuggle banned goods, and enter into all sorts of illicit trade. The profession was ruled by gangs which skimmed off the cream. But if a man was clever, had luck, and built connections, he could in time save enough to start a small shop, or even buy a truck of his own. Once launched, there was no telling how wealthy he might become.

Truck drivers made much more through chicanery than they did through honest labor. One way or another they hoped to make a killing and launch themselves as independent businessmen. They were thus quickly corrupted, and came closer to being *liu mang fan tzu* (rascals) than the *lao shih kung jen* (honest workers) people expected them to be.

Somehow or other Lao Hei had come through this school unscathed. He was too honest and straightforward to consider how he could cheat to become rich. In spite of beatings and insults he had stubbornly persisted in finding out all he could about trucks and engines, hoping he would be able to earn enough to eat.

But even though Lao Hei had not surrendered his innocence, he had unconsciously absorbed many of the prejudices of his trade.

One day there was a tractor that no one could get started. When they called in Lao Hei he was very confident.

"All you have to know are a few tricks of the trade," he said merrily.

Within a few minutes the engine sputtered to life. But when the other students asked him how he had done it, he said gruffly: "It's running now, what difference does it make? The main thing was to get it going, wasn't it?"

His classmates did not agree. They criticized him severely.

" 'Get it going!' Listen to him talk. Whenever we're in trouble we're supposed to call Lao Hei."

"We're here to learn how to fix things like that. Suppose it happens when you aren't around? Are we supposed to quit work? No, you've got to tell us how you did it. That's the only way to serve the people."

"Lao Hei, there's a name for that. That's technical selfishness. The master craftsman wants to make a secret of his skill, wants a monopoly. Those days are gone now. The only way for us to go up is to all go up together."

As Lao Hei listened to their words he grew more and more uncomfortable. His face flushed and his eyes looked uncertainly from face to face.

"I'm wrong," he said finally. "I'm sorry, there should be no secrets among us. I'll show you how I did it, but just remember, nobody told me these things, as I am telling you."

"Of course, that was the old society," said Li Chen-jung, the middle peasant's daughter. "But things are different now. Who wants to go on making the mistakes of the past? We've got to help one another."

Everybody agreed with that.

A few days later Lao Hei again put the wrong foot forward. He and I had been working together all afternoon and became better acquainted than before. Lao Hei said to me:

"Lao Han, let us become blood brothers. I would like to become your brother, and we two would always help each other."

"Sure, Lao Hei," I said. "I'll be your brother." And so we agreed. But when Lao Kuo, the militiaman from the Taihang mountains, heard about it he was upset.

"Lao Hei," he said, "what's this talk of brothers? Are we not all equal here? Should we not all help each other without condition, and if need be sacrifice our lives for one another? We are all class brothers, poor peasants and workers, and we must all stick together. What's the matter with the word *tung chih* [comrade]? Is there any finer relationship between men than that? Blood brothers—that's a feudal idea—to form a small clique, we two against the world. No, Lao Hei, that's a feudal thought."

At first neither Lao Hei nor I understood what Kuo meant, but as time went on we learned to see the problem in a new light.

In the meantime, Lao Hei got married. This was the biggest social event South Ridge had seen in some time and we all made the most of it. The marriage happened very suddenly. One of the tractor students, Liu Mei-sheng, decided that Lao Hei would make a good match for his sister. Since the girl was well educated, had had two years of middle school, could read and write and even do algebra, we were curious as to why Liu had chosen Lao Hei. Although we never got to the bottom of it, one significant factor did come to light. The Liu's had landlord connections. They had probably decided that an alliance with a working man might prove useful in the future—it would consolidate their status in the new society, so to speak—and so Lao Hei and little Lin were introduced. When they met, the young people decided they liked each other's looks and agreed to the match.

For Lao Hei this was a tremendous stroke of fortune. In the old society he could not have thought of marrying so soon. Any bride would have cost him a lot of money. To marry an educated girl would have been out of the question. Once the affair was settled he went around in a sort of daze.

How much the girl actually had to say in the matter was doubtful. Once the family decided this was what they wanted for her, I think she probably did as she was told. She did not have any reasonable alternative. But, she was lucky. They might well have decided on somebody far worse. Lao Hei was strong, handsome, harworking, honest, and kind. In the old society she might have ended up as some corpulent squire's second wife.

To call this a love match, however, would have been to stretch a point. The two only saw each other once before they agreed to get married. Nevertheless, for the Chinese countryside in 1949, it was considered a free marriage. The young couple were not married off sight unseen by their parents as had been the custom for thousands of years. They had at least met and agreed. This was a big step forward in human relations and was celebrated accordingly.

The wedding was simple but gay. We all took the morning off. Lao Hei's mother, a bound-footed old widow in black, came down from Shihchiachuang. The bride and her family arrived in a cart decorated in red. The school fixed up a two-room apartment in one of the compounds for the new couple and the students decorated it with pictures and bright slogans on red paper: "Celebrate the new style marriage!" "The masses have all been liberated!" "Abolish feudal ways where parents rule the marriage and ride about the streets in sedan chairs." "When husband and wife are of one mind, production goes fastest."

In the center of the wall hung a banner beautifully inscribed with these words: "One is a worker, one is a student, together they form a revolutionary couple. The boy is a mechanic, the girl will be a teacher. Both strive to be models. Such workers are the main force of the reconstruction of our society! Tractor Training Class, North China People's Government."

On the brick sleeping platform was a new quilt of flowered cloth, a gift from the government.

The ceremony itself was over quickly. We crowded into the newly constructed dining shed to witness it. Director Li introduced the couple. Then they had to get up and say a few words. The students demanded a blow-by-blow account of their romance, but both Lao Hei and his bride were tongue-tied. Lao Hei stood there grinning foolishly, while the girl looked at the floor and obviously wished she could sink out of sight. Then the bride's brother said a few words. Lao Hei's mother stood up and said how glad she was to see her son married at last, and to such a fine girl. Close friends rose to speak and hinted broadly that what was expected of the couple was production, and not just the mechanical kind either. Roars of laughter greeted these sallies.

Finally the embarrassed, blushing pair were allowed to sit down and we all celebrated by eating peanuts and drinking tea. The young couple held open house in their room the rest of the day for those who wanted to pay their respects, read the slogans, and feel the new quilt.

The wedding was hardly over before family trouble began. Lao Hei's mother, who had been waiting all these years for a daughter-in-law, decided not to return to Shihchiachuang. She moved in with her son and expected little Lin to take care of her. But the bride had other ideas and soon the two started quarreling. Lao Hei, caught in the middle, was unable to do anything with either of them, and came to class completely distraught. He even forgot the Chinese characters his wife so painstakingly taught him each day.

Director Li finally had to take a hand in the matter and persuaded Lao Hei's mother to go back to the city. In the new society, he told her, it is no longer right to expect your daughter-in-law to be a servant. It is better to let the young people work out their own lives. The old lady departed indignantly for Shihchiachuang, but relations between Lao Hei and his wife improved sharply as soon as she had disappeared.

And then Lao Hei disappointed us again. Everyone thought that since the girl had already started middle school, she should have a chance to complete it, and the middle school in the county seat was only a few miles away. With Lao Hei to support her, she could easily enroll there and go on with her studies.

But Lao Hei was stubbornly opposed. He wouldn't listen to reason. He wanted his wife home. Perhaps he thought the gap in their educational levels was already too wide and feared to increase it, or perhaps he was afraid she would take a fancy to someone more polished than he. Or perhaps he just thought, like so many millions of other Chinese men of peasant background, that a woman's place was at home. Whatever the reason, he absolutely refused to let his wife go back to school, so she had to settle down and be a housekeeper.

His fellow students did not let the matter rest. They argued with him whenever the question came up. They told him that this was a feudal tail on his thinking and ought to be cut off. But it was a long time before Lao Hei changed his mind and by that time he was well on the way to becoming an engineer.

Chapter 6

Donkeys, Rifles, and Tractor Engines

ONE MORNING I WENT OUT to the tractor lot and found Lao Hei fanning a blazing fire built under the crankcase of a cold tractor. After the flames had licked the gray iron for some minutes he stepped on the starter. Nothing happened. He then got another tractor to haul his machine around and around. The engine was so stiff that the rear wheels skidded in the dirt. Lao Hei decided that he wasn't hauling it fast enough. He shifted the towing machine into second gear. Still no results, so he called for more weight on the rear wheels. Several students jumped on the drawbar. At last the tractor jerked to life. Immediately Lao Hei opened the throttle wide. He shoved the transmission into high gear and forced the tractor around in tight circles, skidding first one wheel and then the other.

To the uninitiated students all this looked very impressive, but to anyone who felt what was happening inside that tractor it was a demonstration that set the teeth on edge.

"No, Lao Hei," I said. "That's not the way to start a cold machine. You have to be gentle with iron, gentle, gentle. You must ease it into motion, let it warm up slowly, allow the oil to reach every part. Each time you force things you take hours off the working life of the machine."

This plea was a part of the constant struggle we had to wage against the sloppy habits and rule-of-thumb methods brought to the school from China's dusty highways and makeshift repair shops by Lao Hei and the other truck drivers. The class could not have succeeded without these men and their experience, but at the same time I could not allow them to pass on to the rest of the students their improvised solutions, their disregard for grit and dirt, their faith in the hammer and pliers as the basic tools of the mechanic's trade.

China's great tradition of precision craftsmanship had definitely not been inherited by her mechanics. The infinite patience taken by the Chinese with wood, ivory, silk embroidery, and jade was abandoned where iron was concerned. Metal objects were expected to be tough. What did it matter if one beat them, dropped them, or failed to wipe them off? In part this attitude grew out of ignorance, but ignorance could not be blamed for all of it. Much more important were the conditions that prevailed in the trucking industry. Truck owners operated on a shoestring. They supplied neither the buildings nor the equipment for proper maintenance. Hired drivers could hardly be blamed for slackness when the owners themselves did not care enough about their vehicles to build shops and provide tools, parts, and facilities.

A vicious circle developed. Mechanics were careless because their tools were poor. Owners dared not buy good tools because their mechanics were careless. No one trusted anyone else. A good wrench could easily be swapped at a profit for an inferior one. Fine parts could be taken out of the engine and replaced with shoddy imitations. Gasoline could be siphoned from gas tanks, good mineral oil diluted with crankcase refuse or exchanged for the products of seed pressing mills. No one knew exactly what he was buying. Everybody expected to be cheated and everybody cheated in turn.

Under this kind of care new vehicles deteriorated rapidly but old vehicles were kept running for decades. Fifteen- and twenty-year-old trucks were common. They rattled by in second gear, hoods gone, radiators boiling, four out of five lights dead, half the spring leaves broken. A copper strip against the steering column served as a horn button, a heavy chain around one tire served to keep the wall from blowing. An extra man rode the running board to stoke the wood-burning gas producer and wield the crank when the engine stalled. To save fuel the drivers coasted down hills, and to save brake linings they pressed steadily on the horn. The horn was the one absolutely essential accessory. Lights, brakes, window glass, self-starters, all could be dispensed with, but not the horn.

The fact that these trucks ran year after year could not be attributed to any solid mechanical skill, but only to a certain dogged perseverance on the part of the truckers. They had to keep their trucks running because there was nothing with which to replace them—a worn-out truck traveling fifty miles a day still hauled cargo more cheaply than a wheelbarrow. So they worked out many ingenious ways to patch and repair what never should have worn out so rapidly in the first place.

In the realm of patching and tinkering these Chinese drivers had no equal in the world. They knew how to replace the membrane on a fuel pump with a piece of pig's bladder, how to make old spark plugs spark by forcing the high voltage current to leap from wire to plug, how to revive an old coil by baking it in an oven, how to cast parts in a blast furnace made of discarded oil drums.

Nevertheless, for all this ingenuity, they were always behind in the race against wear and disintegration. They were so busy improvising ways to travel a few more miles that they had no time to work out a real system of preventive maintenance. And, of course, they had no funds with which to carry out any proper plan.

Trucks careened through the North China dust without air filters on the carburetors, oil filters on the oil lines, or breather caps on the crankcases. Radiators were filled with alkali water that plugged the blocks with lime. This state of affairs had continued for so long that most people in the trade took it for granted. Truck drivers and mechanics did not even realize the harm that dirt could do. They did not know how to create precision fits, how to measure tolerances, how to prevent unnecessary wear.

Coupled with this ignorance was a certain arrogance. With a rough-and-ready solution for everything they felt no need to study scientific methods. They thought that their ways were good enough. After all, had they not kept trucks on the road year after year? They resented any suggestion that they still had something to learn.

In essence, what these truckmen lacked was a real feel for machines—that sixth sense that comes from growing up with engines, that sense which grasps the machine as a living, breathing, vibrant whole with its own demands and its own set of laws.

I wanted the students to acquire this feeling. I wanted them to learn to love the tractors as a soldier loves the rifle on which his life depends. I reminded them how well rifles actually were kept by the soldiers that passed on the road. Army men, when they had a few spare minutes, polished and oiled every part of their weapons, they always kept a wad of cotton stuffed in the muzzle to keep out dust and damp, and when they put their rifles down or picked them up, they did so with great care, as if the steel were fragile glass.

I also compared the machines to the peasants' draft animals.

"Look how carefully the farmer tends his donkey," I said. "He gives it a warm room in the house, he selects the best fodder he can find, he chops it carefully and watches to see how it meets the donkey's appetite. He makes sure the donkey has cooled down before he offers it water, and he lets it warm up slowly before putting it to heavy work.

"Are our tractors any less deserving? A tractor can plow far more in a day than several dozen donkeys. Well tended, a tractor will outlast any donkey. But it can't protect itself at all. It is completely dependent on you. An abused donkey can always lie down and refuse to move. But an abused tractor cannot retaliate. It cannot fight back.

"Of course a tractor can complain. But you have to learn to understand the language of the tractor when it speaks. You have to pay attention and listen, for the machine has many voices and they all speak at once. You have to learn to tell one from the other."

The students responded enthusiastically to these ideas. They began to listen to the engines as a naturalist listens to bird calls. They learned to cut out one cylinder at a time and analyze the difference in the sound. They took rods and held them against their ears and picked up the tapping of valves, the knocking of bearings, and the humming of the various gears.

From the beginning I tried to get them to do even the smallest things right. Before starting to work we laid out boards on saw-horses and covered them with paper. We placed our tools on the paper, never on the ground. On each nut we used the wrench that fit it. We wiped all parts clean and wrapped them up. We washed them with kerosene before replacing them.

It soon became clear why Chinese mechanics had long since abandoned such practices. Boards were scarce, paper scarcer. Only a few copies of the *People's Daily* arrived each day and these were read and reread by group after group. There were no piles of old papers lying around. Kerosene for washing parts cost the equivalent of $3.00 a gallon and rags were nonexistent. Everyone used worn clothes to make cloth shoes. What remained when the shoes, in their turn, were worn out was good for nothing. For wiping grease and dirt, cotton waste had to be bought from the textile mills of Shihchiachuang. This meant a three-day journey by foot and train. As for wrenches, we had almost none. We scoured the markets in the railroad towns but turned up only a few worn, open-end tools. If it had not been for my little leather case with a set of tools from the hardware stores of Los Angeles, we would have been helpless.

In spite of these difficulties I insisted that cotton waste, newspapers, and kerosene were cheaper than machine parts. Director Li backed me up with an adequate budget, and so the new rules took hold.

A strict table of inspection and maintenance was drawn up and posted at key points around the school. Each study group took responsibility for one tractor and challenged the rest to a contest. Inspection committees were elected to judge the work week by week. They awarded points to those who did the best. As a result the tractors shone as if they had just come off the production line. Not a drop of grease or a streak of dust was allowed to remain.

The students learned the daily maintenance rules by heart and practiced the routine inspection so that it became second nature —first look at the oil stick, then check the radiator, then test the fan belt, then open the air filter, and so on right around the machine. *Pao yang* (maintenance) became the chief slogan of the school and I was renamed Han Pao-yang, or "Maintenance Han."

Once this attitude took hold, anyone who neglected a machine, dropped tools on the ground, failed to cover up parts, or forgot to make a proper inspection was severely criticized in his or her group for lack of a proper "working style."

Faced with this kind of an offensive, even the most blasé truck drivers and mechanics began to change. Lao Hei gave up building fires under tractor engines, and Lao Wang, whose special charge was the school's precious Dodge truck, demanded a set of good tools to replace the pliers and screwdriver he had depended on for so long.

Self-and-mutual criticism helped solve the problem of sloppy technique. It also helped solve much more basic problems of outlook and attitude as the students met every Sunday after supper to criticize weaknesses and help each other become better revolutionaries, better builders of a new world.

Chapter 7

"I Accept Your Criticism"

ONE SUNDAY EVENING I dropped in on one of the mutual aid study groups. Chang Ming moved over and motioned for me to sit down. The students were sitting or lying in the straw listening intently to the speaker, Liu Kai-ming, a strongly built peasant lad who had a veritable passion for basketball.

"I don't know why," Liu said. "I can't concentrate. Whenever I try to sit down and look at my notes I get sleepy. I can't go on."

"You're never sleepy when it comes to playing basketball. You always seem to be wide awake then," said Liu Po-ying. His voice came from a dark corner. There was only one light in the room, a dim flame from a twist of cotton in a bowl of oil. It lit up the faces of those directly around it and cast grotesque moving shadows on the mud walls.

"Yes, that's right," said another. "Morning, noon, and night you're out on the threshing floor bouncing the ball around. If you can't get a game going you play by yourself. Why can't you put the same effort into study? It must be because you don't like to study."

"Well, I only had a few years of school. Studying is hard," said Liu.

"It is just as hard for the rest of us. Many of us never went to school at all. But the people sent us here to learn, not to play basketball. We are eating the people's millet. How can we face them if we don't study hard?" asked Liu Po-ying.

Liu Kai-ming sat for some moments in silence.

"Does anyone else have any suggestions for Comrade Liu?" Li, the group captain, asked.

"My idea is the same," said another voice from the darkness.

69

"Comrade Liu doesn't apply himself. I think it is only laziness. But if he makes an effort he can overcome it. I think we ought to help and remind him from time to time. If he will agree to it, I myself will help him to review the work, once a day, or every other day. We can't just tell him what is wrong. We have to help him."

"That's right," said Li. "How about it, Comrade? Do you want Lao Wang to help you or have you a better idea?"

"I don't know, I never was lazy when it came to field work. But books make my head swim. I think . . ."

"You don't have to decide now. You can talk it over later, but I think it would be a good idea," said Wang.

With that they passed on to the next person.

Sunday evening meetings such as this became the workshops where the new working style was hammered out.

"Working style" meant more than simply the proper way to look after machinery. "Style" reflected a whole background and point of view. Most valued was a dedicated proletarian "style," as distinct from the loose individualism of the peasant or the aristocratic feudal "style" of the landlord.

To call an action "feudal" was to level the most serious criticism, to call it "proletarian" was to give the highest praise. What people meant by a proletarian "working style" included hard work, unconditional devotion to the common task, a spirit of mutual help, concern for public property, prompt attention to problems, precision, cleanliness, and responsibility. These were the working class virtues needed to overcome the basic selfishness, slackness, and diffuseness of village life, where time was measured by how long it took to smoke a pipe or eat a bowl of millet. People believed that without some of the natural discipline and cohesion of the working class, the peasants could never free themselves from feudal bondage. It was, therefore, important for everyone to develop a proletarian style of work.

Self-and-mutual criticism was the method adopted for the cultivation of this new approach. Each person was encouraged to speak frankly of his own work and study and then to ask for criticisms and suggestions from the rest of the group. Criticism was not supposed to tear others down, but to help them overcome shortcomings, to help them change and grow. Instead of forming cliques and gossiping behind each other's backs, the students and staff aired their views in open meetings and strove to reconcile them in a principled way.

They took seriously the words of Mao Tse-tung: "If we have shortcomings, we are not afraid to have them pointed out and criticized, because we serve the people. Anyone, no matter who, may point out our shortcomings. If he is right, we will correct them. If what he proposes will benefit the people, we will act upon it. . . . Our cadres should be concerned about every soldier, and all people in the revolutionary ranks should care for each other and love and help each other."

This was the spirit behind the meetings, even though in the heat of the moment those involved sometimes failed to live up to it. During the first weeks there were frequent quarrels. Students sometimes went to bed angry and embittered. But as the process continued week after week attitudes improved. Those who at first found criticism hard to bear learned that it really was given in a spirit of helpfulness. They found that when they accepted it and tried to overcome their mistakes their peace of mind returned and their relations with others changed rapidly. Resentment gave way to appreciation for the concern shown by others, and the warm comradeship and mutual trust that existed within the group was enlarged and strengthened.

On the Sunday evening that I visited the study group, the main criticism was directed at the group captain himself, Li, the intellectual. He brought up the question of his own short temper.

"I quarreled with Lao Chang the other day. I know that is wrong. I thought, he is always getting in the way. He is so slow. So I spoke to him sharply. That was wrong. I should have explained instead. My trouble is individualism."

"Yes, but you ought to examine it more deeply," said Chi Feng-ying, who was sitting next to the light. The free ends of the towel around her head cast a shadow on the wall that looked like the drooping ears of a sheep. "You are always impatient with people. Even now you want to settle everything right away. You study hard yourself. You are very bright. You have been to the university, so you ought to help others more and patiently explain

things. You ought to consider what is at the root of your thinking."

"Yes," said another. "Sometimes you act just like a landlord.
One would think you thought you were better than other people.
You must realize that your education was made possible by others'
hard work. For every one who studies hundreds must sweat in
the fields. There is no particular merit to being a student. If things
had been the other way around anyone might have done the same
as you. So you should really think about it. Impatience is not just
a part of your character. It has to do with your outlook."

"I have thought about that before," said Li, "but I haven't
overcome it yet. It is true, I thought I am a student because I
am bright. I did not worry about all the hard work that others
put into it. I took it for granted. But I no longer think that way.
I know that there is labor in everything. There is no idleness without labor, there is no studying without labor—just as now, how
could we study if there were nobody to support us, if we didn't
get millet from the government?"

"Lao Li, you speak well," said Chi Feng-ying, "but when you
really understand something it must show in your life. Your
actions must be changed by it. I do not think you have faced the
problem. Not to the bottom anyway."

"Well, I need your help. Everyone should say what they think
and give me their criticism."

"It is better for you to think about it yourself," said Chang
Ming, knocking the charred end off the wick so that the flame
glowed suddenly brighter. "Surely there is pride behind that
impatience. Until you conquer that, it will show up all the time.
It is no use putting on a big hat, 'I am impatient,' 'This is individualism,' or some other ism."

"Are there any other ideas?" asked Li.

"Yes," said Liu Po-ying from the corner. "You should try to
be more objective. You see something and you make up your
mind right away. And then you won't hear of any other opinion
—like that spark plug the other day. You said it must be the
points. We took the distributor apart six times but still had the
same trouble. It turned out to be the plug after all. You should
keep your mind open and listen to others."

"But there was something wrong with the points," said Li.

"No there wasn't. Han Chiaoshih came along and found one of the plugs skipping. He fixed it right away."

"But the points were burned. The tractor wouldn't even start."

"That's just what I mean," said Liu Po-ying. "When you think you are right no one can tell you anything. You won't listen."

"Lao Liu is right," said another voice from the shadows. "You always know best."

Several others spoke up in agreement.

"Let's ask Han Chiaoshih," said Li, very upset. "Was there anything wrong with the points?" he asked, turning to me.

"But don't you see, Comrade Li," Chi Feng-ying put in hastily before I could answer, "it is you we are talking about, not the points. That's just one example. There are lots of others. The other day you said it wouldn't freeze. But it did freeze and Lao Kuo had to get up in the middle of the night to check the tractors. Ours still had water in it. There have been other times. Lao Liu is right. You are too hasty in your judgments. And then you won't admit you are wrong."

"I accept your criticism," said Li, "and I will surely change." But though he said the words, it was clear that in his heart he did not yet agree.

"We raised these opinions to help you, Lao Li," said Chi Feng-ying, "so you must not get upset but think them over. Perhaps there is much truth in what we say."

In this way the students challenged, changed, and strengthened each other. Slowly but steadily everyone's outlook was transformed. Peasants, students, artisans, workers, even the sons and daughters of landlords, who entered the revolutionary ranks learned to forget themselves and think in terms of what was good for the whole.

Chairman Mao had written that it took ten years for an intellectual to overcome his various conceits and illusions and really devote himself wholeheartedly to the service of the people. A laboring peasant could do it in far less time. He had fewer layers of illusion to peel off. The same was true of a worker. But everyone, regardless of origin, had in some degree to slough off the remnants of selfishness, hypocrisy, "face," and mutual antagonism carried over from the past. On that there was general agreement.

I too came to realize the importance of all this when I saw in practice that technical knowledge alone was not enough to create a love for machines or a sense of responsibility for public property. It was those students who most thoroughly identified their interest with that of the Chinese people and who saw the tractors as levers for the liberation of the whole countryside who worked the hardest, took the greatest pains, and assumed the heaviest responsibility.

When, after a period of warm weather, the temperature suddenly fell below freezing, it was not Chang Ming, the ambitious engineer, or Wang, the would-be truck driver, who got up in the middle of the night to check all the tractor radiators and make sure they were drained. It was militiaman Kuo and Youth Corps leader Li. Their concern for public property surpassed their concern for any personal property, such as their private toothbrush or precious padded quilt.

They felt a sense of duty not only to their own people but to the working people of the whole world, for as Director Li said, "These machines were created by American working men and women. They were given to the people of China by the whole world in order to help us overcome the devastation of war. Now the people have entrusted us with their care, and they expect us to use them well and wisely for the benefit of all. Can we let down our own people? Can we let down the people of the whole world by treating the tractors carelessly? No, we must care for them as we would our own children."

Chapter 8

Tractors Need Room to Turn Around

IN JANUARY MY WIFE BERTHA, who had stayed behind in Connecticut when I volunteered for the UNRRA tractor unit, arrived in South Ridge after a long wait in Nationalist Honan and an adventurous trek through the mountains from Hsinhsiang, where a Catholic priest had helped smuggle her through the lines. She immediately found work as the nurse of the training school and ministered to all the ills of both students and staff. She had very little equipment and almost no medical supplies, but her many years' experience made up, in part, for any such lacks and her prestige soon equaled that of any doctor in the West.

Since Bertha was not acclimated to the North China winter, Director Li ordered a stove built beside the *kang* or brick platform on which we slept. The flue of the stove directed heat and smoke under the *kang* and we fired it with coal night and day. Because the wind changed frequently the stove sometimes made no difference to the temperature in the room and sometimes made the *kang* so hot that we could not sleep. Before we had learned how to regulate it, we ran through our whole winter's ration of coal and had to return to living as everyone else did, in a room that was just as hot or cold as the air outside. Padded clothing during the day and extra quilts at night kept us alive.

In February an experienced organizer, Chang Hsing-san (Chang Three Names) arrived in South Ridge. He came directly from the Ministry of Agriculture of the People's Government of North China to take over the farm in Thousand Ching Basin which had hitherto been managed by the local man, Big Yang. The farm was to be called Chiheng because its land ran through two counties—Chi and Hengshui. Since the students being trained under Director Li were destined to become tractor drivers on Chiheng Farm, the training class became, on Chang Hsing-san's arrival, the Chiheng Farm Tractor Brigade. But it was too early to plow, so classes continued as before.

I always remember Manager Chang as bald, but that is not correct. He was not bald, but clean-shaven. Whenever the hair on his round head got long enough to darken the skin with a five o'clock shadow he called the barber to shave it off. This was a common custom in North China.

Chang was a commanding figure, six feet tall and well proportioned. Just as Director Li's clothes always seemed too big for him, Manager Chang's clothes, though made of the same material and equally well padded, always seemed to fit. He wore his garments with the offhand ease of one used to fine cloth and expert tailoring although he had probably rarely enjoyed either. He was a landlord's son who had graduated from Yenching University (China's Harvard), spoke a foreign language (English), and had held responsible positions in various Liberated Areas governments. He came from Tinghsien, a county that lay south of Peking and up against the Wutai mountains.*

From his temperament and bearing I felt that Chang would have done well in the army, but he had chosen a completely different road. At Yenching he studied agriculture. After graduation he worked in government experiment stations and local agri-

---

* Tinghsien became known the world over when Jimmy Yen, Chiang Kai-shek's favorite literacy expert, chose it for an experiment in rural reconstruction. The peasants were organized into co-ops, new breeds of pigs and chickens were imported to raise the level of livelihood, and reading circles were set up in every hamlet.
Why Chiang and the Rockefeller Foundation, which supported Yen's endeavors with cash, should want to spend so much money in Tinghsien had never been clear to me until I talked with Chang, who remembered the pigs with favor. "More revolutionaries and Communists were born and raised in Tinghsien than in any other county in China," he said. "I myself have five brothers. All six of us are in the Party. One is a railroad executive in the Northeast, another is a staff officer in the Northwest, a third is in the Supply Department of the North China forces. Other families from our home area are the same. No matter where you go you'll find Tinghsien cadres making revolution."

cultural bureaus until the Japanese invasion. Then he joined the guerrilla forces in Central Hopei. As the Liberated Areas expanded he was asked to organize rural supply co-ops in the enemy's rear. By the time the Japanese surrendered he had built up an extensive network of stores and supply lines.

Chang was an energetic executive. From the time of his arrival the affairs of the farm began to hum.

When he asked Big Yang how much wasteland was under state control Yang told him: "Probably 20,000 *mou*." "Probably!" Chang exclaimed. "We can't farm probably. Any child knows that the first thing about a farm is the land. How much? Where does it lie? What quality is it? They haven't even made a survey!" Chang was quite upset with Big Yang, who overnight had been demoted to second-in-command.

"When I arrived," Chang told me, "no one even came to the gate to meet me. They didn't have a quilt for me to sleep under. I almost froze. Did you ever hear of such a thing?"

It had not been easy for him to step out of a big regional job and come down to this isolated place. And then to be treated like an intruder when he got there!

The fact of the matter was, of course, that Big Yang resented the demotion. Not only had Chang come down from the North to take over his job, but he began to turn things upside down before he even got in the door. The very first day he wanted to know how much land there was, who had made a survey, whether or not the boundaries had been staked, and a thousand and one other things.

That was Chang's way. He had more than a trace of *chi hsin ping* (anxious heart sickness). He liked to see things done, done fast and done well. Sloppy methods, country ways, didn't suit him at all. And besides, there was no time to waste. If a farm was to get under way a start had to be made right away.

As soon as I could get a tractor fit for work, Manager Chang had Lao Hei drive out and plow a line around the land that was still unclaimed by any peasant. This businesslike act created a stir in the villages around the basin. It brought the whole question of tractors and public land to a head. The people were worried about the future of the land. The fact that there were several thousand acres of wasteland in Thousand Ching Basin was without doubt a great loss to the country. Anyone could recognize that. It should be plowed up and put into crops. On the other hand, the huge tract of idle land was a convenience to the peasants who lived nearby. It made fine pasturage for sheep. When one of them needed a little extra hay for the donkey or cow he could always go out and cut wild grass; such grass, dried and bundled, could always be sold at the county seat on market day. Most important of all, he could always go out and reclaim a little extra land. Although most of the basin had been spoiled by salt brought to the surface by floods and needed washing on a large scale before it could ever be really productive, there were areas of very good soil scattered about. If a peasant had the time and energy he could always take a hoe and add a *mou* or two of wheat to his holdings. Dotted here and there in the basin were patches of cultivated land belonging to many different families.

The peasants wondered what would happen to their economy if the whole basin were plowed. Where would they pasture their sheep? Would there be any place left to cut grass? What would happen to those plots of land in the middle of the waste that they had worked so hard to reclaim? The village chairman of South Ridge came to discuss these questions with Manager Chang. He was a fussy little man who had held office for many years—right through the Japanese occupation, in fact. Elected after the Liberation because of the support he had given the peoples' guerrillas while pretending to be a loyal puppet of Japan, he busied himself about village affairs like an old hen with a brood of young chicks.

The village chairman's talk with Manager Chang began a series of conferences that soon expanded to include the whole neighborhood and lasted for weeks. At the final meeting, the problem of the wasteland was presented to the village chairmen, local labor heroes, and model peasants of the twelve communities most concerned with the future of the basin. This gathering was held at Han Family Village, a low mound in the middle of an area covered with the crumbling adobe dwellings of people who all bore the same surname—Han.

This village was the home of Han Ta-ming, the famous labor

hero of the Chin-Chi-Lu-Yu Border Region whom I had met two years before. He had prospered greatly in the meantime and appeared at the meeting in a brand new set of padded clothes. His face had filled out, he was stouter, but people said he was still the same keen, hardworking team leader that he had been in the past.

Han was very enthusiastic about the tractors. At regional meetings he had talked with people from many places. Leading cadres had told him about machine tractor stations, electricity, and mechanized farms where one man could handle several hundred *mou*. He was eager to see such things come to pass and was ready for new things and new ways.

When Chang Hsing-san appealed to the peasants for cooperation in building a large-scale farm in the basin, he counted on Han to catch the vision and help explain it to the rest. He was not disappointed.

Chang reminded the peasants of the need for utilizing every inch of productive land to support the front and revive the economy.

"Chairman Mao has given us the formula for victory in four lines," he said. " 'Soldiers at the front advance/strengthen revolutionary discipline/raise production one inch/and nothing can defeat us.' We here cannot advance at the front, but we can grasp hold of a very important point—we can raise production, and more than one inch at that. Since the villagers don't have the power to reclaim the land, the government is preparing to do it with machines.

"Comrades, machines need room to turn around. Even a small tractor can plow fifty *mou* in a day. A big one can plow one hundred and fifty, but not if the plots are small. We must give them room. We have to let them go straight. In the Northeast there are places where the driver plows in one direction all morning. Then he eats lunch and plows back in the afternoon. That is efficiency.

"What we propose to do is to swap your small plots in the middle of the basin for land at the side. We will plow the new plots for you. We will plow six inches deep, not three. Then our tractors will have room to turn around. Someday, when we learn to use these machines well, we will help you on your own land. This is a pioneer project. It can change the face of all China and bring an abundant life for everyone. Comrades, what do you think about it?"

When he finished speaking the peasants turned to each other and started talking all at once. When the conversation died down a little, Han Ta-ming spoke up:

"Countrymen, it was not so long ago that we were pulling plows with our bare hands. I still have the welts here on my shoulders made by the rope. Then the *kuei tzu* [devils] were driven out and the land was divided. We got oxen. But life does not stop flowing. The iron oxen have come to replace the flesh-and-blood beasts. We've got to look ahead. New things are happening all the time.

"You wouldn't ask an ox to plow its own stall. You wouldn't try to pluck a chicken with a scythe? I say the comrade is right. The machines need a clear path. I've heard that there is a harvester with a knife as long as a mule cart. Can we expect that knife to jump our plots? No. We have to give the machines a chance."

Some supported Han, but others were skeptical.

"Those plots are the best land. Where is there land so good?" asked one.

"Where will we pasture the sheep?" asked another. "When all the grass is plowed under where can we cut fuel?"

Chang explained that there would still be plenty of land left that was too alkaline for crops. As for the fuel, the plowing would turn up quantities of heavy grass roots, at least a two or three years' supply. By then coal would be coming in on the railroad, or they could think of other means. True, the quality of the land that replaced the odd plots would not be up to the old, but it would be plowed much deeper and would be easier to work. Also it would be nearer home.

Han Ta-ming was enthusiastic. "How long will it be before our group can buy a tractor?" he asked. Eight or ten like-minded peasants joined him in discussing this question. In the end all agreed to return to the villages and urge the people to swap their scattered plots.

Back in South Ridge there was at least one family that could not be persuaded to cooperate. Its members hated the tractors, the tractor class, Chiheng Farm, and everything it stood for. That was the family of landlord Yang (not a relative of Big Yang, the manager). Yang had once owned the compound in which the school was housed. The gutted tower had been his also.

All these dwellings were now "struggle fruits"—confiscated landlord property held by the village for distribution to the landless and homeless. Since South Ridge, like Han Family Village, had suffered heavily in the famine of 1944, there was enough housing to go around for the time being without distributing the Yangs' courtyards, and so the village had loaned them temporarily to the government.

Landlord Yang and his family still occupied a small section of the many buildings that had once been theirs. By blocking several passageways and knocking a hole in the outside wall, they had created a small yard that was completely cut off from the rest of their former holdings but still had access to the street. Yang's son and daughter-in-law lived in another such sanctuary at the very back of the compound. This son wore a slouching felt hat reminiscent of Tientsin waterfront loafers. This gave him a dissolute appearance that was heightened by the pallor of his complexion, the stoop of his shoulders, and the slow movement of his limbs. He was most probably an opium smoker. With the coming of spring he went almost daily to the fields that remained his, but from the expression on his face it was clear that the very idea of labor was obnoxious to him.

Old Yang himself was a hearty man who affected peasant clothes, but the attitude of the family was nevertheless amply revealed by his three-year-old son. This chubby, well-fed baby, instead of playing games and joking with the students like the rest of the village children, always scowled as they went by. Several times he attacked individuals as they walked down the lane by running out and striking at their legs with his fists and shouting. "*Ta ta* [hit, hit]," as he ran. At other times he swore at them under his breath.

One day his father caught him at it and gave him a sound

thrashing, but that didn't make him friendlier. When no one was looking he threw stones at the tractors.

A few days later Hsueh Feng, the man who had accompanied me to South Ridge on the mule cart, found Yang and his eldest son striding around the compound where the tractors were parked.

"What are you doing here?" Hsueh Feng asked.

"Just looking over our property, your mother's . . ."

"What do you mean, your property? This belongs to the village."

"Just the same, it's my property," said the landlord, "and I'll have it back someday too."

"You'll have it back? And who do you think will allow you to steal it from the people?" Hsueh demanded.

"Oh, the day will come. You think the sky will not change again?"

"Get out of here, you turtle's egg. Get out of here this minute," shouted Hsueh.

"Who is going to throw me out?" asked Yang.

"I am, right now," said Hsueh, and advanced toward him ready to strike.

Yang retreated slowly out the gate with his son beside him cursing and swearing all the time.

This incident was reported to the County security forces for investigation but Yang was never arrested.

"What about these ex-landlords?" I asked Hsueh the next day. "What kind of a future do they face?"

"That depends on their own behavior," he said. "They have their land, their houses, their draft animals, and their implements. If they go into the fields and work they can make a fair living—at least as good a living as anyone else. But they aren't allowed to vote, or join the Peasant's Union, or the Women's Association or the Consumer Co-op, or any other popular organization."

"Why is that?"

"For centuries they have dominated the village. They are literate, they are experienced. Given half a chance they would not hesitate to seize control once more, take their lands back, and set up the old system. And if they can't do that they try to disrupt the new system the peasants are building.

"Under the leadership of the Communist Party the peasants have taken away the land on which the landlords' power was based. But this power was also buttressed by custom and tradition, clan ties, temple rites, superstition, classical education, opera, and in a hundred other ways. Now our people are trying to build something new. They want to do away with superstition, clan quarrels, the oppression of women, the domination of the village by cliques and factions for private gain. The landlords want no such thing. They resist and sabotage it. Therefore the peasants have isolated them. They do not allow them any influence in village affairs."

"Will that always be so?" I asked.

"No. If the former landlord works honestly for five years and is law abiding he can be brought back into the community with full rights provided that the people want him."

"Will five years change a landlord's point of view?"

"In some cases. Five years of hard work growing your own crops can influence a person quite a lot. Also, by that time the peasants' power will be well established. New customs and traditions will already have taken root. Everyone's outlook will be different then. It will not be so easy to create disunity. The peasants will have had five years' experience in working together and running their own affairs."

All the other families in South Ridge who held land in the basin agreed to swap lots. Some did it with enthusiasm, some did it with reservations, but none of them flatly opposed it as did Landlord Yang, and so the first concrete steps in the creation of Chiheng Farm, the project that was to launch mechanization in all North China, were taken. It was none too soon in view of the situation in China as a whole. The civil war had reached its climax. Overall victory for the Revolution was in sight and soon problems of reconstruction would replace those of war.

Chapter 9

Battles North and South

THE WINTER OF 1949 was a crucial one. Day after day the news from the front dominated all other concerns.

Starting with the Manchurian campaign in the fall of 1948, the big cities of the North, which had been held by Chiang Kai-shek ever since the Japanese surrender in 1945, fell one by one. Chinchow, Changchun, and Mukden were liberated by November 1. While the armies of General Lin Piao raced south across the Great Wall toward Tientsin and Peking, the armies of Chen Yi, Chen Keng, and one-eyed Liu Po-cheng were locked in a gigantic struggle with Chiang's main forces in the lowlands around Suchow. Close to half a million troops were fighting on each side. This battle, one of the largest ever fought, ended in the complete annihilation of Chiang's forces north of the Yangtze and liberated most of the territory between the Yellow and Yangtze rivers. The attack on Tientsin a few weeks later, and the surrender at Peking of Fu Tso-yi's North China troops, set the stage for the drive across the Yangtze that was the main event of the new year.

This succession of victories kept the school in a constant state of excitement and anticipation. On February 7, the day that the news of Peking's surrender reached South Ridge, the whole village went wild with joy. The students formed a dance brigade and went weaving down the village lanes to the accompaniment of the most joyous rhythm their drummers and cymbal beaters could devise. They were joined by another brigade made up of village children. People came running out of their houses to see what all the noise and commotion was about.

84

"Peking is liberated. General Fu has surrendered," the students shouted.

Peking liberated! . . . Listeners stood still for a moment, letting the news sink in, trying to adjust to all that it meant. Then they rushed indoors to find other family members, to bring them out on the street to join the happy throngs.

It was difficult for a visitor to appreciate all that this news meant to the local peasants. Peking was not just another city, but the ancient capital of China. Its liberation was a sign that the war would soon be over, that their sons and daughters would soon be home, that twelve years of bloodshed and strife, a century of Western intervention, and millenniums of landlord rule would soon come to an end. It also meant the economic revival of a region that had been living on a subsistence economy since 1937.

The peasants of South Hopei lived and labored on a long funnel-like plain where all roads converged on Peking and Tientsin. For more than a decade they had been cut off from the commercial, industrial, and cultural heart of their region. Although they had maintained contacts, smuggled goods, and sent students across the lines, the Japanese and then the Nationalist blockade had grievously disrupted the normal flow of life. This hit the peasants of Hengshui, Szehsien, and Nankung especially hard because they had long been traders as well as farmers. Almost all of them had relatives in the shops and markets of the cities to the north, and tens of thousands had traditionally survived in the winter by peddling goods bought in city markets in the rural areas. Suddenly the area was whole once more. Contacts, trade, and interchange of all kinds would quickly revive. Relatives could be visited, families united, commercial ties renewed.

The elation over the victories in North China was tempered somewhat by the realization that Taiyuan, the great industrial city of Shansi to the west, and Tsingtao, the main commercial port of Shantung to the east, were still held by Nationalist forces, not to mention the vast area of South, Southwest, and Northwest China. The front lines had suddenly shifted to distant places, but the fighting was far from over and there would be many bloody battles before peace could come to the whole land.

That the Nationalists still possessed dangerous strength was brought sharply home to everyone in South Ridge by the sight and sound of American-built transport planes flying day and night from Tsingtao into Taiyuan. Their route lay right across Thousand Ching Basin. The first time one of these planes appeared the students were out practicing driving. Thinking that the aircraft might return to attack them, they quickly dispersed the tractors, covered them with sorghum stalks, and took shelter behind the mounds left on the flat by the salt washers.

But the pilots, flying less than a thousand feet above the plain, ignored the tractors. They were busy freighting the war lord Yen Hsi-shan, his gold bars, his household treasures, his wives, children, concubines, and close relatives out of Taiyuan before the city fell. At the same time they brought load after load of ammunition to its besieged defenders.

The planes made everyone angry. Who could say how much of the toil and sweat of the Shansi people was flown out with every load? How many peasants had been bankrupted, how many girls sold into slavery, how many people reduced to a diet of bark and leaves so that those gold bars could be piled up? Now that accumulated wealth, product of the labor of millions, would be lost to China. The peasants shook their fists at the planes as they flew over.

The disasters at the front threw the Kuomintang into confusion and brought the question of peace to the fore once again. The peace offer Chiang made in January led to headlines all winter.

To make the offer plausible the Nationalists put on a new face. Chiang retired to his summer house in the hills and General Li Tsung-jen suddenly took over the negotiations. But what kind of a compromise was possible at that late date? Director Li gave a talk to the students in which he outlined the terms put forth by Chairman Mao.

Chiang had offered to talk peace on condition that his constitution, his political system, and his armies be preserved. Mao answered with eight points. The bogus Kuomintang constitution would have to be abolished, the bogus political system terminated, the armies reorganized in accordance with democratic principles, war criminals like Chiang and his top generals would have to be punished, the land of the landlords divided, the capital of the four families and the war lords confiscated, all treaties betraying the nation abrogated, and a political consultative conference convened to set up a new government.

"That is our minimum program," Director Li said. "How could we demand any less? Anything else would be betrayal of all that the people have fought for not only during the last three years of civil war but during the eight-year-long anti-Japanese war as well."

Then he repeated an old folk tale which Mao Tse-tung had borrowed from Aesop:

"Once a farmer came across a viper that lay injured and help-less on the ground. The viper appealed to the farmer for mercy. The farmer took pity on the creature, picked him up, put him inside his shirt, and carried him home. The warmth of the farmer's body revived the viper. He sank his poisonous fangs into the farmer's flesh. Within a few days the farmer died. To make peace with those gangsters now without removing their fangs would be like the farmer who took pity on the viper. Defeated and wobbling, they appeal to us for peace, only to turn and rend us when they have recovered a little strength."

This tale moved the students. They discussed it during the newspaper reading period in the morning and whenever they had a free moment during the day. The consensus was that a viper's fangs must certainly be removed. There was no sense compromis-ing with the Nationalists: not until the gentry's power had been broken and the people assured of control could a settlement be made with individuals.

"They should have thought of this when they launched the war in 1946," Chang Ming said. "Then they were riding high. Every-one who disagreed with them was silenced or wiped out. They would not hear of a coalition. Now it is too late. They are on the run. I say we should drive on Nanking and capture Chiang alive."

In the old days the saying was: "Good iron is not fashioned into nails, nor good men into soldiers." The Japanese War and the fight for land had changed all that. Now people considered it an honor to serve in the armed forces and so did their families.

Recruits were no longer called *ping* (soldiers) but "army men" or "People's Fighters," and the whole community took responsibility for them. There were many sons of South Ridge people in the ranks.

When the young men left home, the whole community saw them off with music and dancing. They rode through the village on horseback, with red flowers pinned to their chests, while all the children from the surrounding neighborhood raced down the road in front of them, the old people lined the lanes, and the young students and village cadres marched behind, singing, cheering, and shouting.

By the doors of a soldier's home hung a wooden plaque which read "Glorious Army Family." During the planting and harvesting season these families received special help from their neighbors. The mutual aid teams set aside days to help with soldiers' crops, taxes were lightened for their families, and whatever aid came from the government was allotted to them first.

When the New Year came it was the custom for neighbors, village cadres, students, and government personnel to visit the soldiers' families as a token of gratitude for the sacrifices their sons and daughters were making, to talk with them and find out if they had any problems, and to encourage them to work hard for victory.

The tractor students at South Ridge turned these visits into a major celebration. They made several banners with slogans such as "All people unite to save the country," "Defend the fruits of land reform," "Drive to Nanking and capture Chiang alive," and "Hail brave fighters' families." With drums beating and cymbals clashing they went from door to door.

"Comrade Wang, come out, come out," they shouted as they passed before a plaque-decorated home. "Happy New Year! Happy New Year! Long life to you and your son at the front."

When the old peasant and his wife appeared, blinking in the bright sun and tongue-tied with emotion, a great cheer went up.

"Long live brave fighters' families."

"Long live the People's Liberation Armies."

When the old man had finally gathered himself together to say something, he asked them all in.

"Come in and sit a while."

"Thank you, Comrade, there are too many of us. We came to congratulate you. No need to cause you trouble," and with that they moved on to the next house. Everywhere they met with a warm welcome, and their cheers brought tears to more than one pair of eyes.

The armed forces soon began to appear in the flesh as contingents of Lin Piao's Northeastern legions passed through Chihsien. They were marching to take up positions along the north bank of the Yangtze where the final offensive was being prepared. First several squads came through. A few days later columns of hundreds followed. Behind them the army poured down by the ten of thousands. Soon the whole countryside was alive with troops on the move. Travelers from Shihchiachuang and Tehsien reported that the highways on either side of the plain were crowded too. The armies were moving across the flats of Hopei by three routes, east, west, and center, down to the Yellow River and across it to the plains of Kiangsu, Honan, and Hupei.

None of the highways were big enough to hold the mass of men and carts and weapons. They overflowed onto the side roads, and from them onto the paths and cart tracks between the villages. Looking across the plain one could sometimes see as many as four or five columns moving, some close at hand, some in the middle distance, and others so far away that their red banners could just be glimpsed now and then as they pulled out from in back of one settlement and then faded from sight behind another.

Toward evening platoons chose the villages in which they would spend the night. They found lodging in various peasant homes, but cooked their meals outdoors, each squad around an iron pot. They quickly made friends with the people and were followed everywhere by crowds of children. Here a soldier fixed a little girl's school slate, there another played ball with three laughing boys, over by the well two machine gunners helped an old woman bring up water; wherever possible the army men helped with whatever chores the people had to do.

The three basic principles and eight points for attention that have been the rules of conduct of the Communist-led armies in China since their inception were strictly observed: Pay for what you need. Replace the doors that you sleep on. Don't take as much as a needle or a strand of thread from the people. Don't molest women . . . Simple rules like these had won the trust of the peasants ever since the first irregulars had marched down from Chingkangshan twenty years before. Now they guided an army millions strong and did more to defeat Chiang Kai-shek than all the ammunition and weapons in the world.

In the gathering twilight, as the soldiers sat on bricks and stones and unwound their cloth puttees, people crowded around to ask news of their own sons and daughters. Had they by chance heard of Han Hsu-ming? When last heard from he was with Lin Piao's army outside of Mukden. Did they know Li Su-kang who went into Tientsin with one of the first platoons?

Every once in a while these questions brought answers. The villagers found a fighter who knew one of their own. Or—even more exciting—a local man left his own unit for a few hours to drop by and say hello. Then family and friends stayed up all night listening to tales of the victorious campaign.

In the morning the soldiers got up before dawn, swept the courtyards, replaced what they had borrowed, filled the water jars, and were off before anyone had time to say goodbye.

Several times the tractor students went out to the highway four miles away to watch large detachments pass.

The highway itself was crowded with train after train of mule carts hauling grain, ammunition, and heavy weapons. Interspersed with the carts were cavalry brigades mounted on fat Mongolian ponies, necks arched and long manes flying.

The troops made no effort to look smart, or to march in step. They had come a long way and had a still longer way to go. Five hundred miles to the Yangtze, fifteen hundred to Canton. In their cloth shoes they walked along with an easy swinging gait that took them thirty miles a day. Each man carried his millet rations in a long sock-like bag thrown over one shoulder. On the other shoulder he carried a rifle, while around his waist hung heavy cloth belts full of cartridges. The weapons were all battered Japanese products or shiny new arms just shipped in from America. Everyone said Chiang's Manchurian armies had been very well equipped. The People's Liberation Army had made the most of it.

Each outfit carried its banners in front. The red silk battle flags floated proudly in the wind. Some banners were plain. Others bore the name of the army, the division, the regiment, and told of awards and honors won in various campaigns. Much of the history of the war could be read in those banners. Some units had marched all the way from Yenan in the Northwest to Harbin in the Northeast and back again.

I was surprised to see how many women were marching. They were not combat soldiers but headquarters personnel, medical orderlies and signal operators. They looked healthy, tanned, and strong. There was something about the sight of those marching columns that stirred the blood far more than any dress parade. The shuffle of cloth soles in the dust, stained padded clothes well creased at the joints, the sweat-drenched towels under fatigue caps, the jangling cooking pots, the shining rifle barrels, each with its plug of red cloth in the muzzle, the mules, sweat covered and straining, their harnesses and halters decorated to suit the drivers' taste—all this spoke of battle seasoning, of struggle and hardship, of months and years of fighting and marching. For the men and women in these ranks war had become a way of life.

The students took a tractor with them to the highway. The soldiers called out as they passed: "Comrades, what is that?"

"That is a tractor, an iron ox."

"And can it plow?"

"Sure it can, fifty *mou* in a day."

"Take good care of it. After we've marched to Canton and driven that dog's fart Chiang into the sea, we'll be back to help you," shouted one of the men.

"We'll be here," responded Lao Hei. "Hurry back, hurry back."

Chapter 10

Drama Under a Rising Moon

As THE PEOPLE'S LIBERATION ARMY, one million strong, gathered on the north branch of the Yangtze for the great push into South China, our tractor students wound up their formal studies with a final examination. It took considerable effort to find an examination that did not involve reading, but we did. We took various parts from the tractor, set them out around the courtyard, and tied a number on each one. The students were supposed to walk around and write down the name of the part and its function. Those who could not write went in later with members of the technical group and dictated their answers. The examination was quite an event because no one had ever seen such an examination before. In fact, many had never faced any kind of examination in their lives. But, because of their group study, everyone knew the material and everyone passed. Chang Ming and Li Chen-jung got outstanding marks, but the emphasis was on the standing of the group, not the individual. When the group averages were posted Chang Ming realized that he should have helped his roommates more, for they were near the bottom of the list.

After this came graduation, which was combined with the ceremonies officially setting up Chiheng Farm. The double ceremony was a gala occasion to which were invited all the local peasants, labor heroes, chairmen, and secretaries of the two counties, Communist Party leaders from the regional office, school teachers and merchants from Chihsien, and many others.

The students built a big stage of cedar poles and reed mats and covered it with green boughs, red bunting, and flags. Around the village, in addition to the usual political slogans, they posted tractor reminders such as "Don't forget the twelve rules of Maintenance," "Protect children, drive in first gear," "Watch temperature, oil pressure, and ammeter," "Drain the radiator before you leave," etc. . . . On one long wall characters three feet high proclaimed: "Study Teacher Han's spirit of protecting public property."

Farm Manager Chang, who loved to make speeches, gave the main address. He told why Thousand Ching Basin had been chosen for a farm, why the land had to be unified into large blocks, and how the arrangements with the local people were progressing. He outlined what the future held in store for all of Chinese agriculture once mechanization was successful. He sat down to a tremendous burst of applause. As was the custom in the Liberated Areas, he joined in the applause himself, for the honor was not meant to be for the speaker, but for the ideas expressed, for mechanized agriculture, for the future of China, for the local peasants who had sacrificed personal convenience to make the whole thing possible.

Director Li said a few words, and then, one by one, the various local leaders talked. When the speeches were over, the audience moved out to the fields to see the tractors in operation. Ten of the best drivers put their machines through a sort of dance by weaving in and out across the ground in predetermined patterns which included spinning with one wheel locked. Then they all hitched onto plows and turned a furrow up and back, laying open a strip of soil almost sixty feet wide in a few seconds. The plows were exchanged for cultivators and the new land was thoroughly harrowed. Since that exhausted the implements available, the demonstration ended abruptly.

All this went off perfectly, much to everyone's delight, but when we figured out how much gasoline had been consumed to no purpose except the students' vanity, everyone decided that it never should have been done. In spite of the capture of Tsinan airfield and all its aviation fuel, gasoline was still worth $3.00 a gallon and was scarce. The students decided that if anyone wanted to see the tractors in motion in the future they would have to come around when they were really at work.

The evening centered on music and drama, all prepared by the students. The most brilliant achievement and the one that attracted most attention was a series of colored film strips that were cast upon a large white cloth. Director Li thought up the method. He took a camera, removed the back, and wound a strip of paper on the spools in place of film. Cartoons were drawn on the paper and then the whole strip was oiled. With a strong light shining from behind, the pictures were brilliantly outlined on the screen. As soon as they had seen the effect, quite a few people had made film strips of their own.

Chang Ming made the best ones. He took for protagonist the Chinese cartoon character San Mao. San Mao was a Shanghai waif who was always getting tangled up with the Kuomintang authorities. His original creator had used him to great advantage to expose the corruption and cruelty of Chiang's regime. San Mao means three hairs and that was all this waif had growing on his fine round pate.

Chang Ming's San Mao was not from Shanghai, but was a South Ridge village child who was always getting mixed up with the tractors. The first picture showed San Mao looking longingly around a corner at the tractors in their shed. The next showed curiosity gaining the upper hand; he had already climbed the wheel. In the third he had touched the hot line from the battery and received a shock that knocked him senseless. Undeterred, he got up and looked for some other angle of approach. A few frames later he was shown stealing a ride on the plow as the tractor went to the field. The machine crossed a newly plowed furrow and San Mao was thrown off. A big bump appeared on his head and, strange to say, it was exactly on that bump that his three hairs were growing. And so it went. The village children thought it wonderful.

Another filmstrip told the story of Wang Feng-tou, or Wind Head Wang. A "wind head" is a person who always has to be first and fastest, and this "wind head" got on the tractor and drove like a madman. He drove so fast he went right into a tree. Undaunted, he took out another tractor and drove away without checking the oil. He burnt out all the bearings.

"Never mind, there are plenty of good ones left," said Wind Head Wang and he jumped onto a new tractor only to have it burble to a stop—he had forgotten to add water to the radiator —all the bearings had seized. And so it went. The moral: Don't be a "wind head."

A third film told the story of Chiheng Farm, the acreage, the layout, the crops planned, the difference in performance between a tractor and a mule.

The success of these films was marred only by an expensive accident. Lao Hei, who was in charge of lighting, wired the sealed-beam tractor headlights directly to the generator on a tractor. He no sooner started the engine than the lights burned out. Since they were irreplaceable, Lao Hei was very upset. If anyone had spoken harshly he would have wept. Since the show had to go on, there was nothing to do but borrow some headlights from another tractor and try again, this time making sure that the circuit included the battery.

After the film strips came the drama hour. The most uproarious presentation involved the installation of General Li Tsung-jen as the new president at Nanking. It was written by Li, the intellectual, and portrayed in bold strokes the collapse of the Nationalist regime. President Li was getting nowhere with his fellow countrymen, when, alas, he was called before Uncle Sam (a tall character in a paper top hat played by me) and asked to account for himself.

"You great big useless oaf," yelled Uncle Sam. "I gave you airplanes, tanks, and big guns and you lose them all to the Communists. What the hell is the story?"

"We-we-welll," said Li, "I just can't seem to get the soldiers to fight. Whenever they see the People's Liberation Army they surrender."

"Can't get them to fight! What do you mean?"

"Just that, they go over in droves and take their cannon with them. What shall I do, oh, what shall I do?"

"What shall you do?" retorted Uncle Sam in a fury. "The question is what shall I do?"

But as soon as he said this, the whole question became academic because the Liberation Army soldiers, the guerrillas, and the people came rushing across the stage and overwhelmed the conference. Uncle Sam was knocked off his stool. The tall top hat crashed to the ground. The arrogant foreigner fled in fright while Li grabbed his coattails and looked for a place to hide. He succeeded only in tripping Uncle Sam up. They all fell down in a heap and were captured.

By this time the audience was in an uproar and it took several minutes to quiet the applause.

Another play, directed by Sun, the interpreter, whose secret ambition had always been to go on the stage, told a story called "Four Sisters Quarrel."

An old man was celebrating his birthday. His four daughters all came to bring him presents. The question arose as to who should drink the first cup of wine. The father suggested that she whose husband was doing the most for the war should drink first. The first sister's husband turned out to be a front line scout whose work was as dangerous as could be. The second sister's husband was a machine gunner who mowed down the enemy. The third sister's husband was an officer at headquarters and the fourth sister's husband a cook in a field kitchen. "If it weren't for the scout the army would not know where to go," said the first. "If it weren't for the officer, who would give directions and make plans?" asked the third. "But surely the machine gun is the backbone of all the fighting," said the second. While the fourth asked the unanswerable question: "And where would they all be if it weren't for the man who does the cooking? You won't get anywhere, either at the front or at the rear, without hot food in your belly."

Well, the old man was stuck, so he decided to ask which of the sisters did the most work in the rear. But there again it was a tie: one made shoes, another raised corn, the third wove cloth, and the fourth made clothes. The old man decided they were all heroes' wives and heroic women themselves, so he went out for more cups and all five drank together. Just then a group of soldiers came by on their way to the front. The four women rushed out and gave them all the presents they had brought for the old man.

The people were very pleased with this play too, especially the women, and Lao Sun beamed as the cast took their bows.

Each mutual aid study group then presented something, either a dance, or a skit, or a song, and it was long after midnight before the party finally broke up and the peasants went off to their homes under a rising moon.

Chapter 11

Peking on a Tractor

TRUCKS, TRAINS, MULECARTS, wheelbarrows, camels, donkeys, and number eleven jeeps (the Chinese equivalent of shank's mare) had all carried people to Peking in the past, but ours, I am sure was the first group ever to arrive powered by tractor.

A few days after the graduation celebrations, word came from the North China Ministry of Agriculutre that Director Li, Farm Manager Chang, Technician Han, Bertha, the nurse, and a dozen students and a trailer were needed in the city to help set up a State Farm Management Bureau, lay the groundwork for several new state farms, and buy the supplies, fuel, and parts needed for the spring plowing at South Ridge. Manager Chang decided that an extra tractor would not be refused. We hitched the tractor to the trailer and set off for the north.

The trip was a joyful one. Seated comfortably on a mountain of baggage, we rolled through the countryside at ten miles an hour and had plenty of time to see the landscape, sing, and talk. Members of the technical group took turns driving and we pushed right through to Peking—two days and one night without a break.

As we went north we bucked the flow of the armies that were still marching south. The uniformed men and women laughed with delight when they saw the tractor. They seemed to know right away what it was and called out, "*Tuo la chi lai liao, tuo la chi lai liao* [a tractor is coming, a tractor is coming]."

We passed through the heart of Central Hopei, the area that had suffered the most from the mopping-up campaigns of the Japanese war. The villages in the area had survived only because of the tunnel warfare developed by the peasants. Gaping holes in the fields showed where the tunnels had run that once connected every village.

This was the region where Manager Chang had spent the war years and as we rolled up the road he kept recognizing places where he had lived, worked, or fought. At one spot he pointed east to a mud-walled village. "See that place," he said. "That was our headquarters. The Japanese devils held that other village to the north—less than three miles away. We had to be on guard all the time. Our sentries patroled the sorghum and kept a close watch. Whenever the Japanese made the slightest move we took to the fields through the tunnels. We lived opposite them like that for months."

A little later we came to an irrigation canal that wound for miles through the countryside. This large-scale engineering work had been built by the peasants of the area only a year or two before, at the height of the civil war. While Nationalist General Fu Tso-yi's troops were blowing dykes on the grand canal to flood hundreds of thousands of *mou*, the peasants of Central Hopei were digging an irrigation channel to bring water to perennially parched land. Chang had had a hand in organizing that project too, and talked about it with pride.

Toward evening we passed Szehsien and the largest mission compound I had ever seen. This was the center of the famous Szehsien spy case: French Catholic fathers had been charged with operating an intelligence net and reporting to Peking by radio. The suspects were tried and expelled from the area, but the mission remained. The French must have lived like kings. Their church grounds and outbuildings covered an area larger than many a walled county town.

It was still early in the spring and darkness brought intense cold. Shivering, we pulled into Hochien, famous wartime capital of the Liberated Area of Central Hopei. In the blackness the town appeared to be nothing but rubble—great crumbling walls, abandoned courtyards, fallen roofs—but in the center of the destruction we found a couple of prosperous looking restaurants and had a supper of hot soup and noodles. This warmed us up and we started on again. At dawn we found ourselves in the center of the Paoting-Peking-Tientsin triangle, the area where the fighting of the last three years had been most bitter. The effects of the battle were visible everywhere. Hardly a wall still stood. The villages had been smashed back into the mud and dust from which they had originally been fashioned. The Kuomintang general in charge of this area had been a renegade Communist who knew guerrilla warfare better than most Nationalist commanders. He had split his armies into roving bands and carried the fight to the outlying hamlets with utter ruthlessness.

In the afternoon we went through a long stretch of sand where the trucks of the southbound armies were moving with difficulty. We saw big Burma-road Dodges and GMC six-by-sixes, beautifully kept up and expertly driven. I noted particularly how their army drivers double clutched in the sand and pulled through it without either spinning their wheels or letting them come to a stop. They had a degree of proficiency which I had not expected. "If you can do as well with the tractors as those soldiers do with their trucks, our farms will succeed, at least on the technical end," I said to the students.

The sun was still high when we rolled into Peking. Crowds of people stopped to stare as the tractor and trailer roared by. Sunburned and covered with dust, we looked like wanderers from the Gobi Desert.

Peking again!

I had left two years before at the height of the student protest against the civil war. Thousands had marched that spring in defiance of a national ban on street demonstrations, and the unity between the students and the population at large had been such that the Peking authorities had retreated temporarily. On May 25, 1947, the police stayed off the streets and soldiers were confined to their barracks.

I had never seen anything to compare with the energy, organization, and flair for publicity shown by Peking's students during the demonstration that followed. First came the massed columns of marchers holding banners aloft, singing defiant revolutionary songs and shouting slogans in unison. Their mood was neither grim nor violent, but hopeful, friendly, and passionate in demand-

ing a better world. On both sides of the marchers the propa-
gandists went to work. Most numerous were the boys and girls
with chalk. They wrote slogans on every convenient surface: the
pavement, the sidewalks, the walls (Peking has more miles of
blank wall than any city in the world), the arches, the gates, the
doors, the store windows, the awnings, and on all moving things—
cars, streetcars, trucks, even rickshaws. It was as if a flock of birds
had descended and had been joined by an army of literate beavers.
"The people want peace," "The people want to live," "Stop civil
war," "Chinese must love Chinese," "No more hunger"—these
and many other slogans were written wherever there was room
to scrawl a few ideographs. After the chalkmen came the paint-
pot wielders. They painted the same slogans, but larger and with
paint that could not be brushed off. Some used a few lines to
depict starving children, the dove of peace, or empty rice bowls.
Following the painters and competing with them in energy and
determination came the tar-pot artists. They dipped twisted pieces
of cloth into pots of liquid tar and rubbed slogans on the walls.
One group of five girls was especially diligent and before the
march was over they were splashed with sticky tar from head to
foot. In between all these came the pasters. They pasted up a pro-
fusion of paper slogans and posters that often blotted out the
efforts of those who preceded them. Some of the larger posters
were very effective cartoons. People went in for harsh cartooning
then, somewhat in the style of the Russian anti-Hitler work, and
they drew fat reactionary monsters crushing skin-and-bone
people under foot or reclined on the backs of starving peasants
while bombs exploded in the background.

The pasters, the tar wielders, the paint-pot brigades, and the
chalkers were only a part of the effort, however. For there were
also leaflet distributors and newspaper vendors, and, most effective
of all, speakers. The latter stopped and talked to anyone who
would listen. Over here a young man addressed a group of rick-
shaw men. Over there a girl talked quickly and earnestly to the
occupants of a streetcar temporarily stalled in the traffic. The
passengers made a fine captive audience for they were jammed
together, unable to move. As the speaker finished the crowd
clapped and cheered. A young student who spoke broken English

came up to me. "Sir, we are students demonstrating against the
civil war. The government must stop this war. The people are
starving while Chinese kill Chinese. We hope your country will
not send any more arms and will help us build democracy. Please
write all your friends and tell them what we say." A few minutes
later another came and said the same thing in a different way.
They were not angry at me for being an American. They only
asked for understanding and support.

The response of the Peking crowds was disappointing to the
students. They had hoped that thousands would join their parade.
They wanted it to grow into a huge mass demonstration. Nothing
like this happened, but the people were friendly. Many drivers
stopped their cars long enough for the slogans to be chalked
and the posters pasted. Many shopkeepers looked on approvingly
as tar ideographs were brushed on their awning mats. Everyone
bought the papers, even soldiers, officers, and American-trained
flyers. There was no mistaking where the sympathies of the public
lay.

When the march was over the whole route was littered with
leaflets, posters, slogans in red paint, black tar, and white chalk.
It was hard to believe that a few hundred students could cover
such a great area in so little time. But even more remarkable was
the scene the next day. Every single slogan had disappeared. All
the chalk had been rubbed off, all the posters torn down, all the
fresh paint painted over, and all the tar smudged out with black.
Householders were held responsible for what appeared on their
walls and the police had worked all night to see that they cleaned
them up. As the sun rose, I saw an old man rubbing a red wall
with a broom. Beside him stood a policeman holding a gun. Thus
did Chiang restore law and order and erase from sight the truth
that had burst forth so suddenly and beautifully.

Even larger demonstrations had been planned for June 2, 1947.
Students were to march, workers to strike, businessmen and intellectuals to meet. But this time the government clamped down.
An early curfew cleared the streets in all major cities. Troops and
police came out in force. Campuses were barricaded with barbed
wire and tanks were stationed at strategic points. The students
had to call off their march, but this did not satisfy the Peking

garrison command. Hired thugs, directed by police informers, roamed the streets in gangs carrying clubs of wood and iron. As soon as they spotted a student he was ordered to halt. If he ran, shots were fired. If he stopped, the thugs descended on him and beat him to the ground. As many as thirty or forty attacked at once. When they had beaten their victim senseless they picked him up and threw him in a waiting car to be carried who knew where. Many students were never seen again.

Such was the Peking I had left in 1947. How peaceful, relaxed, and yet how full of energy the city was now. There was an entirely different tone to everyday life. Instead of flashy Western suits and loud ties, government workers wore blue padded uniforms and cloth shoes just as they did in the mountains. Airplane heads (permanent waves) and long split gowns had gone out of style. Most girls had already bobbed their hair or braided it, and wore pants like the men. Even shopkeepers and businessmen were catching onto the spirit of the times. Some already went about in Sun Yat-sen-style uniforms. The countryside had taken over the city.

And the police, how their attitude had changed! They no longer yelled or swore at people. Nor did they beat them up. White megaphones had replaced guns and nightsticks. Jay walkers, light jumpers, and careless drivers were politely warned to mend their ways. It was hard to believe that these were the same corrupt, graft-ridden bullies who a few months ago had lorded it over the population.

Now they called everyone "comrade." A pedicab collided with a bicycle. Both overturned and the two men jumped up to curse one another. "Comrade," said the policeman to the pedicab driver, "if you had waited until I gave the signal the bicycle would have passed by. We have to learn to wait our turn. The rules apply to everybody. What a mess our streets would be if they didn't. Next time try not to shove in so fast." And to the bicyclist he said: "Comrade, you too are at fault. You were peddling too fast." Taken aback by this calm approach both men looked ashamed. They righted their vehicles, apologized to each other, shook hands, and rode off. In the old days the collision could have developed into a fist fight. One or both participants might well have been arrested by the policeman and released only on payment of a fat bribe.

The *glacis* around the Legation Quarter had also changed. Ever since the Boxer Rebellion in 1900 the whole area had been cleared, at the insistence of the foreign powers, to create an open field of fire for the machine gunners behind embassy walls. Now, fifty years later, the ground had been reconquered by China and had spawned, overnight, the busiest market in town. Restoration of sovereignty over the area and the march of the Liberation Army up the main street of the Legation Quarter, where for decades no Chinese troops had been allowed, had made a lasting impression on the people of Peking. Two months later, they were still talking about it.

Ironically, the main items for sale on this once forbidden ground were the remnants of army and PX stores left behind by the Marines—canned goods, air-force sleeping bags and parkas, GI boots, mosquito repellent, and Marine Corps knives.

We found the Ministry of Agriculture right in the heart of the theater district, next door to the great Tungan produce market and only a few blocks from Wangfuching, the busy shopping street. The buildings left much to be desired. Built by the Japanese, the two-story monstrosity of kaoliang stalks and plaster had several dilapidated compounds and courtyards. Bertha and I were assigned a ground floor room which contained a wooden bed, the first we had seen in a long time. In the middle of the night the bed crashed to the floor and woke the whole compound.

It is hard to describe what we felt those first few days in the city. The happy crowds, the overflowing shops, the endless streams of pedicabs and bicycles, the busy markets, the palaces, the parks, the red flags flying—all this under a clear blue sky and a warm spring sun. A city that belonged to its citizens at last, a capital and a countryside united as one and the dark days of the Kuomintang gone forever! It went to our heads like wine.

Old cadres like Chang, Li, and Hsueh behaved like children. The first thing they wanted to do was to see the city's famous sights. "After all, it belongs to us now," they said, as if it were some new toy. One might have thought that old hands like Chang, who had lived in Peking for many years and who had graduated from college there, would have been less excited than the tractor students, some of whom had never seen a city at all. But if anything it was the old cadres who were most deeply moved. For twenty years they had lived through the most terrible trials in the countryside. Now at last they had returned as victors to the places they had known and loved.

We went, of course, to the Forbidden City, the palace of the Manchu Emperors, and tried to see all of its 999 rooms in one afternoon. We really had no chance to see anything because Chang and Li were too excited to slow down. When I stopped to look at some old sword or tapestry, Chang rushed me to the next courtyard to see where the Empress used to sleep. When Director Li got interested in the Empress's yellow bed, Chang hustled on to the steam-heated conference room. (They said the steam heat, installed by a Western firm as a gift to Her Majesty, never did work well.) And so it went. We dashed from exhibit to exhibit, and from courtyard to courtyard, laughing, calling to each other, and rushing each other forward as if afraid that the whole thing would vanish. When we were finally shooed out by the guards at closing time, the only things I could remember were a monstrous gilded clock with elaborate moving figures and chirping birds that the Emperor of Austria had sent to China in the 1890's, and a large two-handed sword that only a giant could wield.

On Sunday we climbed into an open truck and went out to the Summer Palace. A lot of other trucks, charcoal burners smoking, were headed the same way. They were loaded with singing school children, government workers from various ministries and bureaus, contingents of factory workers, and soldiers. It seemed as if the whole city was going west. We were unable to hire any kind of boat. There were long lines standing at the reservation booths. We contented ourselves with tickets on the big ferry that plied between the stone boat and the island in the middle of the lake. Then we climbed the mountain that was built from material scooped from the lake bed and drank some soda at the pavilion on top. It was the first soda any of us had had for years. It didn't taste very good—lukewarm and sickeningly sweet—but

soda was one of the benefits of civilization and since we had just
come out of the back country we were in no mood to forego it.

Bertha and I were overwhelmed. Between bouts of sightseeing
and celebrating, we managed to get a glimpse of the future. South
Ridge was no longer going to be central to the development of
mechanized agriculture. It was to be the site of just one of many
state farms. Only enough drivers would stay to handle the land;
the rest would be sent to new places like the South Airfield at
Peking, to the steppes of northern Chahar, across the Great Wall
from Kalgan, and to Lutai, an irrigated rice farm on the Pohai
Gulf north of Tientsin. Farms were also planned for Yungnien
near Hantan, where the land was shaped like a doughnut around
the town moat, and Poai just north of the Yellow River, where
the Taihang mountains began to crowd the plain. And that was
only the beginning. Reports of wasteland were coming in from
every province in the North. Trained people were needed, hundreds of them. A national training center would probably be set
up near Peking and a special bureau established to administer the
farms.

Plans and projects were developing with astonishing speed.
Peking had only been liberated a few months. South China was
still in Kuomintang hands, but reconstruction was underway in
the North and was rapidly gaining momentum.

How recent the Liberation was was illustrated by the watch the
soldiers kept on the city at night, especially in areas where there
were government buildings. After ten or eleven it was always
best to walk down the very center of the street so that whatever
light there was shone fully on you, for the sentries, who were
combat troops on temporary guard duty, challenged everyone
who passed with a swift:

"Shui [who]?"

The proper answer was "Wo [I]" and a continued steady pace
forward. But sometimes the challenge was so unexpected and so
sharp that we forgot what we should say and stopped dead still.
This alarmed the sentries and they came out to look us over.
When they saw our faded blue jackets and visor caps they waved
us on.

Chapter 12

Liberated Tientsin

THE NEW NORTH CHINA MINISTRY OF AGRICULTURE decided to send Manager Chang, Hsueh Feng, and me back to South Ridge to carry through the spring plowing. Since Bertha was not feeling well—she had an attack of dysentery and a slight fever—I persuaded her to stay behind at the Foreign Language School where there were other British and American teachers to talk to.

Chang, Hsueh, and I returned to South Hopei by way of Tientsin. There we hoped to buy oil and take possession of some grain drills that had been left behind by AMOMO, UNRRA's successor in the farm machinery business in Nationalist China.

Tientsin was busy and looked much as it had when I left it in 1947. One conspicuous difference was the absence of United States Marines. The shops and bazaars were filled with goods and the people looked prosperous. Liberation by the forces from the countryside had not had as much influence there as it had in Peking: Tientsin had had fewer government cadres to start with and a lot more commerce and industry. The city had always had a more metropolitan flavor.

Down along the docks where we went to get oil I overheard several workers discussing the Communist cadres.

"Look at these new cadres, they are all right," said one.

"Yes. It is quite a change. Remember those Kuomintang bastards. Each one of them had two or three concubines all wrapped in fur coats."

"These fellows bring their wives with them."

"Yes, and the women wear the same clothes as the men. They all have a lot of children too."

"Even the men wash their own clothes."

"They are ordinary people like ourselves."

"How would you like to have the others back?"

"*Hu shuo pa tao* [are you crazy]?"

This conversation took place a few blocks from the railroad yards where I had spent so many hours in 1947 trying to get the UNRRA tractors loaded and shipped off to the interior. Having nothing else to do I had watched the "Battle of the Coal," the relentless struggle between the dust-blackened, rag-covered children of Tientsin and the Nationalist railroad guards. The children went after coal wherever they could find it: they scraped it still hot from the grates of waiting engines. They pushed it lump by lump down the drainpipes of the tenders. They picked it from cars on the sidings, and they gathered it from between the rails and alongside the tracks.

The soldiers had been there to stop the children. They had guns and they used them, not usually to kill or to wound, but to frighten.

The soldiers had kept a little fire burning off at one side, and they sat around it boiling tea, looking neither to the right nor the left. The children crept into the yard, crawled through the fence, crouched under the standing cars, scampered across the tracks, and hid behind the iron wheels.

They were lean, ragged little boys and girls not yet in their teens, shoeless and hatless, sharp eyed and nimble footed. They came by the tens, the twenties, and the fifties, hundreds of small invaders sneaking softly into the yards and disappearing in the maze of waiting trains.

When they thought the area was safe, the children came out, first two, then four, then eight, twenty, fifty—all laden with coal. Some carried lumps heavier than their own bodies or bags taller than themselves. They ran across the open tracks, ducked under waiting freight trains, scrambled toward the yard fence, now only sixty feet away.

Suddenly the soldiers came running. The children were cut off. They fell back, dodged behind cars, under axles, behind wheels. But the soldiers hunted them out and ran them down. The children were smart and dispersed in many directions. Some got away. Others, unable to escape with their coal, dumped it out

of the bags, dropped their big lumps, and ran for safety unen-
cumbered. A rifle barked. Then all was still. No children could
be seen anywhere. Here and there in the yard little piles of black
coal lay shining under the noonday sun.

Then the soldiers summoned their own gang, a gang of small
girls, to fetch the coal. The girls picked up the lumps, scooped
the abandoned piles into bags, and hauled the coal off to a large
heap that already contained several tons.

I had been wondering, until that moment, why the soldiers
took their guard duty so seriously. They ran fast, crawled in and
out among the cars as if their lives depended on it, and repeated
this at intervals all day long and far into the night. Yet they
seemed never to tire. They pursued the little yard waifs like
angry devils hour after hour. Such loyalty among conscript
guards was unheard of. That central pile of coal clarified the
situation. It belonged to the soldiers. The coal stolen by the children from the cars was stolen from them in turn by the guards
who sold it for money, a very profitable sideline for men on
tedious yard duty. How the children must have hated them!

Now the yards contained only trains. No children, no guards,
no battle over coal. This seemingly endless struggle, like so many
others, had been brought to a sudden end by the victory of the
Revolution.

We were buying oil from the State Oil Company which had
taken over the stock of Shell and Esso. There seemed to be some
confusion as to just which oil was which and what each kind was
for. Clerks had a lot of samples and a lot of names but no one
knew what the names meant. I finally chose a sample that looked
like SAE #30 but when I got to the warehouse no one had any
idea where to find the oil that would match the sample. When
we got back to the central office, no one could remember where
the sample had been drawn from so we started all over again.
The officials of the company began to think that I was in-
ordinately fussy and I thought they were quite disorganized.
We even had a few sharp words, but in the end Chang smoothed
things over and we got what we wanted.

Chang advised the State Oil Company to set up a laboratory

to analyze their stock and classify it according to suitable use. Without such a classification they would surely sell the wrong oil to some unsuspecting customer and do a lot of damage. They said they already had the same idea.

While waiting around for the oil I saw my first Chinese-made movie. It was only a short but it was excellent. It told the story of a Liberation Army detachment that billeted for the night in a small village on the Manchurian plain. One of the soldiers accidentally fired off his rifle while cleaning it. The bullet killed a small boy. For the death of the villager the soldier should, by law, have been shot.

The issue was taken to the entire village by means of an outdoor public trial. The rules of the army were read aloud and it was made clear that the soldier merited the death penalty for criminal negligence. Some of the men in the unit rose and condemned him roundly for carelessness. They spoke of the good relations they had always maintained with the people.

Then various peasants got up to speak. They were all opposed to the death penalty. Sentencing the second boy would not bring back the first. After all, was he not a good fighter for the people? They thought he should be given another chance. Finally the dead boy's father took the floor. True, he said, he mourned the loss of his own boy, but it would not help a bit to punish the soldier. Then he made a radical suggestion. "Since I have lost my own boy, and since the young soldier is an orphan, I would like to adopt him as my own if he is willing. Then I would have a son in the Liberation Army."

This suggestion was enthusiastically approved by everyone including the unfortunate soldier. And so it was decided.

The old man and his wife were beside themselves with joy. As the unit marched off the next day the father was shown rushing out to the road with an armful of white steamed bread for his new son to take along. For a long time he stood at the edge of the village and waved to the receding marchers, while they waved back.

It was a simple story but a powerful one, especially when viewed in the light of traditional Chinese militarism. It was portrayed with restraint and realism. The actors were not professionals but real soldiers and real peasants: the film was a documentary rather than a studio production. I felt, as I sat in the theater, that I had actually marched into the village, stood among the people at that meeting, and marched off with the men in the end. It looked like a good beginning for a newborn movie industry.

The Tientsin Agricultural Bureau was an organization set up to take over the huge areas of government-owned land that grew rice—the only important tracts of rice in North China. The biggest of these tracts was at Hsiaochan not far from the ocean where 300,000 *mou* had been irrigated by canals dug by Imperial troops more than fifty years earlier. The Empress Dowager had settled her palace guard, a crack Anhwei army unit, there. They reclaimed large areas of alkaline waste by washing, and planted rice. Their descendants still farmed the area but did not own the land. The land had always belonged to the government.

Under the Kuomintang and the Japanese these fertile paddies had been a rich political plum. The annual rents amounted to an enormous quantity of rice. The right to collect these rents was sublet by various politicos to intermediaries, or second-class landlords, who sometimes further sublet the rights to still another set of intermediaries or third-class landlords. As a result the peasants who actually planted the land were reduced to extreme poverty and often ended up eating the beancake which had been loaned to them for fertilizer by various eager banks.

The Japanese found this system in existence near Tientsin and immediately proceeded to expand it. They opened up huge additional tracts on both sides of the river and leased them out to tenants for rice culture only. This provided both a good source of rice for the Japanese armies in North China and yielded a handsome profit.

The Agricultural Bureau had inherited this vast empire with its three-tiered exploitation system, its irrigation and fertilizer experts, its accountants and clerks and hangers on, its offices in downtown Tientsin, and its warehouses bulging with last year's crops. There was a big cleanup job ahead. Wang, the man in charge, was an old college classmate of Chang Hsing-san. They had studied agriculture together at Yenching and both had gone from there to the countryside. Wang had a nickname—Wang Ta-pao, which meant "Big Gun" Wang. He was called "Big Gun" because of his booming voice and his set opinions. This stocky, broad-faced man with a shock of jet black hair that fell over his forehead had suddenly been thrust into the chair of a big business executive. He seemed to take enthusiastically to the role and found real pleasure in showing us, the country cousins, around the offices, warehouses, and multi-storied dormitories of his rice empire. Still, he looked a little out of place sitting at a big desk with a buzzer on one side and a telephone on the other, a doorman who brought tea, and a shining, twelve-year-old Hupmobile complete with chauffeur waiting at the door.

Wang and the lovely young wife who had just borne him a child lived in one room in the bureau's second hostel. This was a shabby house just off Tientsin's busiest street, Roosevelt Road—one of the few streets in the city to retain its original name.

Wang's situation was made to order for the silver bullets of corruption aimed by the city merchants against cadres just in from the country. But it looked as if Big Gun had the situation under control. A campaign to remove the first and second tier of landlords had already begun.

When we had solved the oil problem, at least for the time being, Manager Chang and I returned to South Ridge by truck, leaving Hsueh Feng to wait for the railroad cars that would move the supplies. For the next six weeks we helped to carry out Chiheng Farm's first plowing.

Chapter 13

Plowing in Thousand Ching Basin

WITH NO BARRIERS OR PEASANT PLOTS to stop them, the tractors, spread out over thousands of acres, moved up and down the Thousand Ching Basin like a swarm of gray beetles. They sounded like a squadron of bombers passing overhead in the distance. Behind each plow the dark earth turned up and over to the staccato sound of snapping roots. Some of these roots were larger than a man's thumb and could stop a tractor dead if the driver did not raise the plow in time. As the new furrows fell snugly into place beside the soil already laid over, the smell of fresh earth, cool and moist, hung for a moment in the air. But the sun soon worked its way with the clods and baked them hard and dry.

One could tell when the tractors had been running well by the density of the groups of village children who were out behind the plows picking up roots for domestic fuel. When tractors broke down the children tended to cluster in large groups behind those machines that were still moving. They pushed and shoved each other for a chance at the fattest gleanings. But as soon as the breakdowns were repaired and all fifteen machines were busily at work, the children were unable to keep up with them and scattered from one end of the field to the other. When this happened, it was obvious that we had had a successful afternoon.

Such results did not come easily that first spring. The students, now suddenly converted into working drivers, still had a lot to learn. Although they knew what made the wheels go round and followed the rules carefully in caring for the Fords, they had by no means mastered them. They had not yet learned to understand the language spoken by pistons, valves, and bearings. They could not hear when trouble was brewing and thus could not prevent it. There was little that I could do about it any more: at that point experience was the best teacher.

Every breakdown was a minor tragedy. We had no spare parts, almost no tools, only one barrel of light oil and a few cans of grease. First the fan belts gave way. Even when properly adjusted they cracked and came apart. We sewed them, we patched them, we tried various substitutes, and in the end had to abandon several tractors because we had no belts for them. A few days later the water pump bearings started to go. The first one broke with a smash that sent the fan into the radiator. We repaired the radiator with solder but we could not solder the shaft or revive the bearings. We had to start cannibalizing.

Cannibalizing in a tractor park is like an accelerating epidemic. To keep one machine running you remove a part from another. The next time something breaks, another part is snatched away, and then another and another until only a crippled hulk remains. Then the disease spreads. Soon there are two hulks, then three, four . . .

I was opposed to this on principle but there was no way out. We had to render some tractors useless in order to keep the rest going. By putting sound parts together we managed to keep fifteen of the twenty tractors in the field.

As water pump bearings continued to fail we tried to rejuvenate them by packing in new grease. We became expert at determining the precise degree of wobble that could be tolerated without disaster. Then generators began to overheat, coils quit sparking, plugs cracked, thermostats in the radiator hoses broke. The cannibalism went on. But so did the plowing. And suddenly we found that the plow points were worn out. A week of breaking that hard soil was all a chilled steel plowshare could stand.

I advised Manager Chang to find a blacksmith and for once luck was with us. In a village less than five *li* away, he found a young man who had worked in Peking factories and was familiar with modern forging methods. Together blacksmith Liu and I learned how to sharpen shares and temper them again. From that day on the village rang with the sound of hammering. We had no sooner solved the problem of the shares than the furrow wheel springs began to give way. But this too blacksmith Liu remedied with new springs stripped from wrecked trucks and sold in the market at Shihchiachuang.

The men of the technical group were organized into a flying repair squad. Each had a bicycle and a set of tools. To make up these sets we divided all the tools on the farm. They roamed the fields looking for trouble, or, to be more exact, rushing from one trouble spot to the next. When they found a problem they could not solve they sent for me.

Weeks passed in a nightmare rush. The days blurred into each other. And so did the sharp outlines of the peoples' characters, for the work dominated everything and fused everyone together into a team that had only one thought—to keep the tractors running and finish the plowing. The whole farm was put on two shifts. We rose at odd hours, ate with our heads still heavy with sleep, worked, ate, and fell to sleep exhausted again, plagued, even in our dreams, by frayed fan belts, wobbling water pumps, and stalled tractors. And yet, as June surged into July, a little order emerged from chaos. More and more often we achieved periods when, for hours on end, nothing went wrong.

On one such afternoon I sat in the middle of the work area waiting for something to break or stop. The roar of the engines receded and grew and then receded again as first this machine and then that one drew away, approached, and then drew away once more. I tried to watch all the tractors at once and to notice whether the plowing was even, whether the soil stuck to the shares, whether anyone had the throttle pulled open too far. Every nerve was taut. I listened to the throb of each motor as if it were my own heart. It seemed impossible for steel and iron to stand the strain of that tremendous speed much longer. Inside the motor blocks shafts, rods, and bearings whirled around several thousand times a minute. Surely something must break. And yet minutes passed, the sun sank lower, and the unbroken waves of sound continued.

I watched the different drivers going by. Li Chen-jung sat

straight as rod, like a novice on a skittery horse. She had no feel for the machine at all, but she was determined to master it. With every muscle tense, she grasped the wheel with one hand and the hydraulic control lever with the other. She kept her eyes on the furrow ahead and saw neither to the right nor to the left. She had no way of knowing what the plow was doing behind her, because whenever she made an effort to look around the tractor swerved.

Chang Ming too, was stiff, though more at ease than Li Chen-jung. He whistled under his breath as the tractor roared along.

Liu Po-ying, a husky peasant, sat in the seat as if he belonged there. As he came down the far side of the field, I noticed that he was sitting half sideways so that he could look both ways easily. But he was so absorbed in the plowing that he didn't even notice the uneven sound of his engine.

"Hey, Lao Liu," I shouted, "don't you hear anything wrong with that motor of yours?"

He stopped and looked surprised.

"The engine? I didn't hear anything."

A quick test showed that one plug was not firing. We cleaned the plug off, set the gap, and replaced it. He went off with all four cylinders firing smoothly.

But troubles never came singly. I had no sooner straightened out Liu Po-ying than No. 10, No. 3, and No. 18 all stopped. One had a plugged gas line, another a plow that wouldn't raise, the third a generator that had ceased to charge. A half hour of feverish work got them moving again and I had a chance to sit down.

Just then Manager Chang came striding across the plowed land, stooping every now and then to pick up a handful of dirt and examine it with evident concern.

"How do you like the looks of that dirt?" I asked him.

"I'm afraid it won't do too well," he said. "I thought that with all this grass it would be full of organic matter, but it isn't. And there are too many salty patches."

"What can we do about them?"

"That's what I'm worrying about. There is a plant that thrives on alkali and also makes good fertilizer. Maybe we can plant a lot of it around, chop it, and plow it under. But that will take years."

"Anyway," I said, "the plowing is going well. The students have done well today."

"Yes," said Chang. And then, as he surveyed the basin with one glance: "That is quite a sight."

As we stood there on that endless flat under the huge canopy of the sky, the machines breaking the abandoned land all around us and the sun slowly falling toward the rim of the world, a sense of deep contentment, a sudden awareness of the richness of life swept over us like a warm wind. We did not say anything, we just stood and watched and listened and let the full excitement and wonder of this new work sink in. Where would it lead us, and these young drivers, and China and the world? This was not the first time that I had seen the tractors go into action on the North China plain, but this was the first time they had really settled in, really come to grips with Chinese agriculture. From this place and through the efforts of these very people forty centuries of farming would be transformed. Could there be any doubt of it?

"You know," Manager Chang said finally, "we'll have to begin thinking of livestock. How can crops be grown on a scale such as this without livestock? We'll have to raise dairy cattle." Yes, he too was dreaming. A vision of the farm as it must someday be was already forming. The great expanse of flat broken by windbreaks, an irrigation system bringing water from the Yellow River hundreds of miles away. Fields of wheat and cotton alternating with wide meadows of clover. Cows, sheep, and pigs grazing in deep alfalfa, and on the high ground great barns rearing their bulk against the sky. All this was not so hard to see in the mind's eye.

It began to get dark. The roar of the engines ceased. It was time to change shifts, to grease and gas up for the night work. We went in to supper. In the dimly lit courtyard we squatted down with the drivers and gulped the hot millet as if breaking a three-day fast. All around grease-stained, dust-begrimed youngsters who had been raw students only a few weeks before ate with the appetites of laboring men.

No one said very much. They were all too tired and hungry. They only shoveled the yellow grain into their mouths, chewed, swallowed, and filled their bowls once more.

On the flat the motors sprang to life again. Lights came on and began their steady motion north and south. I climbed to the top of the landlord's tower, which the farm had already repaired and from which one could see the whole sweep of the basin. The lights of the tractors floated like will-o'-the-wisps over a dark sea.

It was easy to see why the peasants in far away villages had mistaken these lights for dancing fairies and fox spirits when the night shift first began. To quiet their alarm the county chairman's office had had to send messengers throughout the district. They assured the old folk and the superstitious ones that it was not fox spirits but the iron oxen of the Ministry of Agriculture working at night to get the land plowed before the summer rains set in.

Chapter 14

Sugar-Coated Bullets

BY THE TIME THE LATE JULY RAINS began to fall, 12,000 *mou* of
wasteland had been plowed. All the gasoline on the farm had
been used. Breakdowns had reached such proportions that cannibalism no longer solved the problem. A rest was imperative.

The tractors were driven or towed back to the shed and jacked
up. Buyers were sent off to Tientsin to get parts while the
drivers began a series of meetings to study the lessons of the
spring work. Now that practical work had begun, such meetings
were considered even more essential than they had been during
the study period. The object of the meetings was to go over the
whole six weeks of plowing, discover mistakes, analyze the reasons
for them, and take steps to correct them. "Do not fear mistakes,
fear only repeating them," was the slogan.

Those who had done well were praised, those who had been
lazy or careless were criticized. During the course of these meetings the Communist Party branch of Chiheng Farm decided to
recruit three new members into its ranks. The three were Liu
Po-ying, Tung Hsiao-ping, and Liu Cheng-tsung.

Before they could become candidates for membership these
three had to go before a mass meeting of the whole farm and
listen to criticism and suggestions. Serious objections on the part
of the rank and file could prevent their acceptance into the
Party.

Liu Po-ying and Tung Hsiao-ping were steady, hardworking
drivers who had done their utmost ever since coming to South
Ridge. The other drivers had nothing but praise for them. In
regard to Liu Cheng-tsung, however, there were differing
opinions.

Of the seventy students who graduated from the training class, Liu Cheng-tsung alone had been picked to train as a truck driver. This was a special honor. Manager Chang chose him because he wanted someone of strong character and principles to counter the dissolute, devil-may-care tradition which seemed to have claimed almost everyone connected with transport under the old regime.

While the rest of the young people were out on the flat plowing under the hot sun Liu was on the highway with truck driver Wang rolling from one town to the next. They traveled to Weihsien, Hantan on the Peking-Hankow Line, Hengshui, Linching, and Tehsien. They hauled seed, supplies, lumber, cement, and iron.

Although he was of peasant origin, travel was not new to Liu. He came to South Ridge from the People's Liberation Army where he had served for several years as a bodyguard for a leading member of the Communist Central Committee. As a guard he had traveled widely, met well-known people, listened in on important policy meetings, and learned a great deal about the world. He had also learned to read and write and had studied the writings of Mao Tse-tung. He was intelligent, alert, and anxious to serve. If he had any fault it was a touch of vanity. This was common among the bodyguards of leading men, for the latter treated their guards like sons and tended to spoil them. The work was light, the life exciting. It was easy to forget how ordinary soldiers and peasants lived and toiled.

As truck driver Wang's assistant, Liu was in a difficult position. He had to live on good terms with Wang, be a good pupil, obey orders, and at the same time set an example to the older man in honesty, frugal living, and service.

Liu tried hard, but rumors kept coming back from the towns through which they passed that Chiheng drivers ate well, sometimes took a drink, and occasionally stayed in private inns instead of the government hostels where lodging was free.

When Liu got up to speak about his work he mentioned none of this but criticized himself for being slow at learning: "I have been on the road for several weeks already but I have not yet learned how to handle the truck. I cannot shift gears smoothly. The other day I carelessly backed into a building and broke one of the tailboards. I haven't enough regard for the people's property.

"I have also quarreled with Lao Wang because I thought he did not give me a chance to learn. During the whole time he has only let me drive for a few hours. But I should be more patient. There are a lot of things to learn about the truck besides holding the wheel and shifting gears."

"This really is no criticism at all," said Chiao, who had gone with the truck on one trip to help load some scrap iron. "You say you haven't learned fast enough, but then it turns out that it is Lao Wang's fault anyway. I think you ought to speak about living on the road. During this period, when the country is going through such difficulties, we all ought to live in a simple, frugal way. But when the truck got to Weihsien we passed by several good restaurants until we came to a fancy one where there were other trucks parked. We went in and sat down with other drivers and everyone ordered expensive dishes. Chopped cakes with sliced meat, noodles with eggs, even meat dumplings. I ordered a couple of *man tou* (steam bread), but I was the only one. Some of the drivers ordered wine. They were matching fingers for drinks and making a lot of noise. And outside sat our truck with Chiheng State Farm written right across the hood in gold letters . . ."

Before he had a chance to finish, Kang, the buyer, jumped up and launched into a speech on the same subject. "I have often been in Wei County. I have spoken to many people there. The peasants say: 'Nothing but the best for the State Farm. Up at Chiheng they live in real style. They can teach anybody how to spend the people's millet.' I say our truck drivers are giving the farm a bad name."

"Right," said Lao Ying, the tinsmith. "It is no harder to drive a truck than to sit on a tractor all day. Just because truck drivers are out on the road is no reason why they have to have a special diet. There are simple things for sale in restaurants. In the cab of the truck you are not even in the sun. There is a cool breeze blowing, so why should you want meat and dumplings when everyone else is eating millet?"

"You make it sound very pleasant," said Lao Hei, who had driven trucks himself for many years and had also been out on the road for Chiheng Farm. "There is a lot of trouble on the road. Things break down. Then you have to walk miles and miles to get to some town. You have to sleep outdoors and guard the truck and work without food or sleep to repair it. It's not like a game of basketball. More than once I have had to drink water out of the radiator because there was no other water to be found. And in the rainy season the roads are nothing but mud holes. You get stuck all the time and have to battle every inch of the way. There is no off shift—now we'll eat—and no on shift—now we'll work. You have to get where you are going."

For a moment it looked as if an argument would develop over the difficulties of trucking as compared to tractor driving but Manager Chang put the discussion back on the track.

"Every work has its difficulties," he said. "There is no point arguing that. Such an argument can never be settled. The question of simple living, of a proletarian style of work, is something else again. What do you have to say about that?" he asked, turning to Liu Cheng-tsung.

"Well," said Liu, trying hard to remember, "we did usually get together with other drivers and we bought good food. I didn't pay much attention to it. I ate what Lao Wang ate. Everyone seemed to take it for granted."

"But why didn't you speak out? Do you have to follow somebody, even when it is extravagant and wasteful?" asked Li Chen-jung.

"I was Lao Wang's apprentice," said Liu.

"You talk as if this were the old society—apprentice and master. But that is not true. You are equal even if one is learning and the other is teaching. You can raise criticisms. You can refuse to join the banquet. What has that got to do with driving?"

"What good is a Party member who doesn't stand up for the truth?" asked former militiaman Kuo. "A Party member has to have backbone and courage to speak out even if he is all alone."

"This question is an important one," said Manager Chang, "because we all represent the People's Government. If we do not set a good example of hard work and service to the people, who will? Right now when we are on the threshold of victory we must steel ourselves against new temptations. The sugar-coated bullets of the merchants and manufacturers are as dangerous to us as the steel and lead of the enemy. Soon many of us will be going to the cities to study, to buy, to plan. If we cannot maintain our integrity, if we cannot resist the corruption of city life, our whole revolution can easily collapse.

"To maintain our working style in the face of such difficulties is much harder than to plow all day in the sun, or to drive all night through the mud. Up to now Comrade Liu didn't pay enough attention to these matters. He was thinking about learning to drive and he let Comrade Wang set the style in everything. But it is just because he is young, and because he has been raised in the revolutionary army under the leadership of our Communist Party that we expect him to stand firm for what is right, what is fitting and honest.

"Our farm is here to show the people a new way. If it is successful perhaps in the future our whole country can eat dumplings every day. But that time is a long way off. In the meantime we must live as the peasants do and set a good example. We must justify and maintain their faith in us. This is especially important for our truck drivers because they represent us all over the province. They must carry our honest and simple way of living with them wherever they go, so that people will know that we are serious hardworking people who are doing our best for the future of China."

After Manager Chang stopped speaking nobody said anything. Manager Chang lit a cigarette. Liu looked at the ground as if he were thinking hard. Finally he looked up and began to speak.

"It is true. I did not take a forthright stand. I thought, 'It is better to get along with Lao Wang and learn to drive.' But that was wrong. I have done our work much harm. Progress cannot be made without a struggle. I will certainly do better in the future. We have always lived simply here at South Ridge. Certainly we can do the same wherever we travel. I pledge myself to do that."

Everyone thought Liu Cheng-tsung really meant what he said. His candidacy for membership in the Communist Party was approved.

Chapter 15

Trains, Shops, and a Pacifist

IN LATE JULY, a few days after the plowing ended, Manager
Chang and I went to Peking for a second visit. Chang was to take
over the leadership of the new State Farm Management Bureau.
I went along to advise him on technical problems until work
should begin again in the basin. A student, Ma Lien-hsiang, went
with us. He was bound for the tractor brigade at Peking's South
Airfield. On our departure Big Yang, the local man, once again
took over the management of Chiheng Farm.

Before leaving we three travelers were guests of honor at a
wonderful farewell party. The farm cook suddenly blossomed
forth with all sorts of fine pork and chicken dishes which no one
had suspected he could make. In between courses cadres and
tractor drivers wandered from table to table challenging everyone
to empty a cupful of the best Hengshui wine. Chang and I were
the object of a veritable bottoms-up offensive and drank enough
to feel quite dizzy.

It was a warm send-off and we needed it because the weather
outside was terrible. The rainy season had already begun. The
roads were a foot deep in mud. Even the two strongest mules
could hardly pull the lightly loaded cart across the flat. The mud
kept riding up and jamming the wheels. We had to wade around
in water up to our knees and dig the wheels out with our hands.

At Hengshui we caught the night train to Techou. We found
seats in a large roomy chair car and settled down for the night.
What I remember best about the trip was the program the train
crew put on for the enlightenment of the passengers. The conductor and the brakeman took turns shouting through a megaphone. They not only called out the stations, reminded passengers

not to forget their luggage, not to lean out the window or hold children outside the train for toilet functions, but they also gave short talks and lectures between stops. The main topic of the evening was sanitary childbirth. The conductor said that most infant deaths could be avoided by taking simple sanitary precautions. The midwife should wash her hands, the scissors should be sterilized with boiling water, and only clean boiled clothes should be used. Everyone listened very intently and fell into discussion when the conductor passed on to the next car for a repeat performance. Since most of the passengers were peasants it was a topic of considerable concern to them. Countless village babies died of blood poisoning and tetanus at birth because the country people did not understand anything about bacteria. As the train chugged back and forth to Techou, it became a sort of school on wheels. The passengers, with nothing else to do, listened eagerly, and the train crew, anxious to serve them, had time to explain many important matters.

At Techou we had to wait many hours. We hired a room at a tumbled-down inn and tried to sleep but the bedbugs made rest fitful. The town had suffered heavily during the war. The station had been bombed to rubble and beyond the freight yards along the westbound track more than a score of wounded locomotives lay half buried in the fields. They looked like stricken hippos, their great sides ripped open by bomb fragments and peppered with bullet holes.

The station was crowded with people from every part of the plain, peasants on their way to the city, soldiers in transit, cadres from various training schools on their way to take up new jobs. It finally turned out that the passenger train would not be in until the next afternoon because the rains had washed out several big bridges down the line, so we all climbed into a train of freight cars bound for Tientsin. The cars were jammed. Most of the passengers were peasants from the cotton country through which we had been passing. They were on their way to Kalgan and northern Chahar with handspun, homewoven cloth.

They told us that in the North cloth was selling well, any kind of cloth, and that one could buy cattle very cheaply in return.

With transportation available for the first time in many years, machine-made textiles were flooding the plain and undercutting homemade cloth. But across the mountains in Chahar, where no cotton was grown and the people wore clothes made of sheepskin, homespun was still very much in demand.

The cloth these men carried was lovely to look at and very long lasting. Many bolts were fashioned of natural brown cotton or of pinstriped brown and white. Because the threads were hand-spun there was a roughness and variation in the weave that lent it a rustic charm. In America it would sell on the luxury market at very high prices; here it was considered old fashioned and crude.

Each of the peddlers carried only fifty to seventy pounds of cloth with him. They bought the cheapest rail tickets and took along bags of steamed bread and corn cakes to eat during the ten day round trip.

I sat with three men who were all from the same village.

"Why do you all go?" I asked. "Why not send one man off with all the cloth, save the other fares, and save the time of two men, and share the profits?"

This suggestion did not meet with favor. The men looked at each other, and then back at me as if at a loss for words.

Finally one of them said: "With one ticket we can only take one bundle. If we took more we'd have to buy another ticket."

"Even so you'd save two men's time."

To that they had no answer. I decided that they had not reached the point in their village where they would trust one another to take such a trip. This was typical of peasant individualism. It was each man for himself.

"In some villages the co-op handles the sale of homespun," one of them said. "We don't have a co-op yet."

It was exciting to travel up the line on a train after all I had seen on this roadbed during the UNRRA days. It had been destroyed from Tangkuantun to Tsinan during the three years of the civil war. The rails had been buried and the ties burned. Huge ditches had been dug in the embankment, first from one side and then from the other, to make repair work as difficult as possible.

I had wondered if the railroad ever would be restored. Yet here, only a few months after the liberation of the region, we glided smoothly over that same roadbed.

The station buildings were still in ruins, but at each stop merchants and stall keepers were doing a thriving business in hastily thrown up mat sheds. Each railroad center had overnight become a thriving vortex of life and commerce. The communities readjusted as if trade had never been interrupted.

We spent a night in Tientsin. The city was brilliantly lit and the streets were crowded with shoppers. We found a dazzlingly clean shop that sold ice cream, a dozen flavors, and treated ourselves to several scoops apiece.

As we stood on the street watching the crowds go by, a young girl and her boyfriend passed in front of us. They were holding hands. Ma Lien-hsiang was shocked. This was his first trip to the city. He had never seen people holding hands in public before. Such an open display would never be tolerated in the villages of the Taihang. He thought surely the girl must be a prostitute. Why else the waved hair and the lipstick? We had a hard time convincing him that ordinary working people dressed that way in the big cities and that holding hands was a legitimate sign of affection.

Lao Ma was going through the same experience as tens of thousands of other country cadres. They had years of the most intense social experience behind them—war, revolution, guerrilla fighting, land reform, the organization of peasant unions and village councils. In judging the people and the events they knew they were extremely sophisticated. But when faced for the first time with city life, they appeared naive. All sorts of stories went around about the take-over teams that didn't know what toilets were for, or how to turn on an electric light. But the honesty of the rural cadres, their willingness to learn, and their hard work soon won them the respect of the city population.

When I finally got back to Peking I found that Bertha already had a permanent job at the language school and a big room in the former Japanese barracks in the embassy quarter.

I spent the next few weeks buying parts and equipment for Chiheng and for several new farms in Peking and Tientsin. The

wealth of tools and parts in Peking made all of us feel as if we
had stumbled into paradise. Along the north wall of the Temple
of Heaven I discovered a wonderful hardware market. The long
winding street was packed from one end to the other with stalls
selling second-hand tools and parts. Taps, dies, cylinder gauges,
fine micrometers, and microscopes mingled there with old toilet
bowls, worn tires, pink and blue bathtubs, pitchforks from upstate New York, and baby-carriage wheels.

I went there first to buy canvas. I found a little shop run by an
old couple, or to be more accurate, occupied by an old couple
and run by the old lady. They specialized in taking American
army tents apart and selling the canvas in strips. I bought enough
to make covers for all the tractors. There was no fixed price on
anything. I had to bargain and hoped that I knew more about
values than the old lady, which, of course, was unlikely. Quite
accidentally I came across a brand new Ford tractor coil lying
innocently in a nearby stall. It had undoubtedly been stolen
several years before from the UNRRA supplies bound for
Suiyuan. I tried to hide my excitement and pick it up cheaply.
But the wily merchant sensed the excitement and asked a fantastic
price. Although it was worth its weight in gold to us, I left it
there, knowing that no one else could ever use it. Each time I
went back we dickered over it and he finally agreed to sell for
half his original figure. It was still not cheap.

Tientsin boasted an even more extensive market. Instead of a
single street, a whole city block of twisting alleys was packed
from one end to the other with shops. Some of them specialized
in pressure gauges, some in chain, others had everything imagin-
able: wrenches, bolts, nuts, pipe fittings. None of them were
more than eight feet square. The owners lived in the middle of
their inventory. They put boards across stools and unrolled their
bedding in the shop at night. In the daytime they sat out front,
sorted and resorted their wares, and tried to interest everyone
who passed in a sale.

It was tantalizing to see the fine tools, the beautiful welding
outfits, the precision instruments. The most prosperous of the
merchants were worth many thousands of dollars. Most of their
stock was left-over American goods. Since, due to the blockade,

there were no new imports, prices were high and tending higher. The merchants held us up for everything they could.

While having lunch at the Victoria Restaurant, on one of these Tientsin shopping expeditions, I ran into Ann Liu, who had once worked in the UNRRA Tientsin office. She said she was working for the Standard Oil Company. Through her I learned that her sister Mary, also a former UNRRA employee, was employed at Jardin Mathieson's, the big British import-export firm of Opium War fame. I dropped in to see Mary. She was very excited. She invited me to come to dinner and see her father and mother. They lived in a prosperous-looking two-story apartment in one of the former foreign concessions. Old Mr. Liu was well known in the city. He was an early member of the Kuomintang Party and a close follower of Sun Yat-sen. When Chiang Kai-shek took over the Kuomintang and joined the foreign police in slaughtering the Shanghai workers' battalions, Liu broke with him and went into retirement. Now he was a member of the Tientsin City People's Political Consultative Council and met with people from all walks of life to determine how the city should be run.

This family was a study in contrasts. It was typical of many well-to-do families of the post-World War I generation. The older girls, Mary and Ann, were educated in Tientsin's American School, spoke English with an American accent, and read and wrote it much better than they did Chinese. All their lives they had read American magazines, followed American fashions, and tried to ape Hollywood in their dress and manner. In every way they were more like sophisticated Bostonians or Philadelphians than they were Tientsinese. On the strength of their language ability they had been able to get good jobs with foreign firms and international agencies.

The younger girls, on the other hand (there were four sisters and two brothers in the family), grew up after the family fortune had begun to wane, when the Japanese occupied the treaty ports and the whole Chinese people had risen up to fight for survival. The younger girls went to Chinese schools, wore Chinese clothes, and were moved by the currents of thought and feeling that stirred China. All their ties were to their own people. The third sister, instead of learning English and taking a job with some foreign firm, had joined the revolutionary student movement when she was in college, and had gone, with thousands of other students, to the Liberated Areas. She had been at North China University in Chengting when I was teaching English there and told me that she had seen me buying peanuts on the street.

From Chengting she made a very important trip back to Tientsin to see her father. Her father was a cousin and close friend of Fu Tso-yi, the general in command of the Kuomintang's North China armies. At the time of Tientsin's collapse before the Northeastern People's Liberation Army, Fu had 500,000 troops at his disposal, half of them centered around Peking. If he had decided to defend the city a bloody and destructive battle would have been inevitable. It would have resulted in untold damage to the people and the destruction of irreplaceable architectural monuments. If he could be persuaded to surrender peacefully, the war in the North would be over. A precedent would be set for other hold-out armies. Countless lives would be saved.

Mary's sister made a dangerous trip back across the lines to persuade her father to go and see Fu Tso-yi and urge him to surrender peacefully. The old man made the journey. Within a few days the surrender was arranged. One quarter of a million men laid down their arms without a fight.

Just how important Liu's mission was is hard to say. General Fu's army was doomed in any case. To fight would have meant sure defeat, for he was surrounded on all sides. He had been listed as a major war criminal. In return for his surrender he was offered, instead of punishment, a high position in the new government. But Liu, at the least, added one more voice to those urging surrender and thus helped to remove a tremendous obstacle standing in the way of the Revolution.

What made Liu decide to see Fu Tso-yi and urge a peaceful settlement? I thought perhaps he was a revolutionary at heart who saw that China's salvation lay in throwing out the corrupt Chiang regime and putting into practice Sun Yat-sen's three principles. But as I talked to him it turned out that he was a pacifist who was sick at heart and grieved at the interminable war going on in China. He thought that Mao Tse-tung could bring peace and urged his cousin to surrender without a fight in the interest of peace for the whole country.

Old Mr. Liu showed me a manuscript he had been working on for many years. It was on pacifism. He was opposed to war, all war. He did not agree with the Communist leaders that only through a war of national liberation, a people's war against the landlords, the compradores, and the foreign powers who held China in bondage could peace be won. Nevertheless, he did his share to bring about victory for the popular forces and for that he was highly honored by the government and by the people of Tientsin.

To depend entirely on left-over parts found in the markets of Tientsin and Peking would have eventually led to disaster. The Ministry of Agriculture took over the Farm Implement Factory outside Peking's east wall and began to make whatever could not be bought. We went there first for plowshares. With the aid of the factory technicians we worked out methods for making cold cast iron shares that lasted much longer than the steel shares that had originally come with the tractors. They were brittle, but there were very few stones in the North China Plain. The cast iron shares usually lasted without breaking until they were worn out. Then they were melted down for recasting.

The Farm Implement Factory was growing rapidly. New buildings were under construction. All the forges, steam hammers, lathes, and other machine tools were going full blast making irrigation pumps and new type horse-drawn plows. The place had certainly changed since the day in 1947 when I visited it with Skuce, the UNRRA deep-well expert. Then the Kuomintang man in charge had shown us around almost empty buildings. He had talked at length about what he would do when he got new machinery and equipment from America. In the meantime half the equipment that he already had stood idle. In the whole plant we had seen only a few dozen workers, mostly blacksmiths desultorily beating straight some twisted iron. Unfinished pumps, started by the Japanese, lay rusting in the yard. The place was, if not dead, at least dying.

With the Liberation it had sprung to life, as had industry and commerce elsewhere. No one waited any longer for equipment from abroad. As a matter of fact, many valuable machines from America were found stored in crates in the warehouse. The workers immediately put every machine and tool available into operation and started new farm tools rolling off the assembly lines in ever increasing numbers. A new furnace was built to keep up with the demand for cast iron. Repair parts for the huge paper mill across the road were turned out and that too went into production. Both plants were alive with workers, messengers, and engineers rushing from shop to shop. It was as if a winter ice jam had suddenly thawed and set the currents of production free.

What was the basis for the thaw? The land reform. Now that the peasants had land there was, all at once, a huge new market in the countryside for farm tools, pumps, fertilizer, and insecticide. Not far from the Farm Implement Factory a new insecticide plant was under construction, while in the city a branch plant made hand pumps and sprayers. The factories could not keep up with demand, nor was there any hope of doing so. After hundreds of years of stagnation, the countryside had started on the road to mechanization and change. With the formation of land pooling co-ops, hand sprayers and one-man plows would soon give way to full sets of horse-drawn implements, and these in turn would be replaced by tractors and power machines of all kinds. No wonder the workers hurried.

The needs of China's expanding agriculture pressed the workers in these plants almost as hard as the end of summer pressed the tractor drivers at South Ridge. They had plowed 10,000 acres of wasteland and now had to get it sown to wheat—the first attempt at mechanized planting ever undertaken in North China.

Chapter 16

Autumn Sowing

BY THE MIDDLE OF OCTOBER the breath of winter already hung on the air. For days on end clouds filled the sky. Strong winds blew down from the west and the geese, flying southward in enormous wedges, battled to stay on course. Then suddenly the clouds blew out to sea. We looked up at a clear blue sky that went on to infinity, transparent and bright in every direction. In the early morning, after the dew had cleaned the air, we could see all the way to the Taihang mountains, a row of wolf fangs on the distant horizon. But the sun was small in that vast expanse and the days did not get warm even at noon.

The village fields were already beginning to look as barren as the wastelands of Thousand Ching Basin. The tall sorghum called kaoliang that grew rank and lean like something out of a primeval fen had long since been cut down. The millet, too, with its plump drooping heads, had been carted to the threshing floors. Only some ragged corn stalks and a few wisps of unpicked cotton remained in the fields where beans were stacked to dry. The earth, the bare brown earth, which for a few weeks had been smothered under a luxuriant mantle of green growth, now dominated the countryside once more.

The peasants were busy at the threshing floors rolling out the millet with stone rollers, throwing the chaff into the air and watching it float downwind as the yellow grain fell back. Piles of millet, glistening, golden—the very wellspring of life—grew and multiplied. The village chickens would not stay away even though the peasants stationed their children to drive them off with whips. The many-colored birds gathered at the edge of the beaten ground and darted in to pick up stray grains as soon as the children looked the other way.

The roads resounded with the creaking and clattering of iron-tired peasant carts hauling public grain to the railroad station at Hengshui. There a field of enormous reed-mat silos sprang up like a rash of plump toadstools.

In doorways and courtyards village women were padding winter clothing with new cotton. They made clothes for themselves and their children and uniforms for the men in the army. As people donned their new padded suits they suddenly appeared larger, stouter, and, for a few days at least, well scrubbed and clean.

On the flat east of Han Family Village, militiaman Kuo's brigade was busy seeding the last of the winter wheat. It was late in the season for planting wheat but it was unavoidable. As I sat on the trailer that carried the seed and watched the two bright green grain drills move across the fallow land I thought of all the troubles and delays of the last few months.

First the rains had come. They so stimulated the growth of the grass that it soon covered the basin with a mass of green that reached higher than a man's waist. Within a few weeks the spring plowing was obliterated. Manager Yang ordered the land re-plowed. But the supply department was not prepared for such duplication of effort and after a week the tractors ran out of oil. A man was sent to Shihchiachuang to see what he could pick up. He came back with a barrel of "engine" oil. The oil was dark, but appeared to be of good quality. Lao Hei, who had risen to the position of head mechanic, said to go ahead and use it. Within a few days bearings started to fail. Soon all the tractors were laid up with seized rods. Black gum stuck to the pistons, crankshafts, and cylinder walls. The oil, it turned out, was designed for use on the working parts of locomotives. The engines referred to on the label were steam engines.

This mass breakdown was a major tragedy. For a while it looked as if the whole year's work would come to naught. Everyone made melancholy self-criticisms and vowed never again to be careless about oil. A plea for emergency help was sent to

Peking. Parts, mechanics, and a technician named Chiao Shih-ju were rushed south by truck.

The disaster turned out to be less serious than had at first been imagined. When all the gum was scraped away most parts were found to be in working order. Enough tractors were renovated to resume the plowing.

Then the supply department ran out of gasoline. Gasoline was being shipped up river from Tientsin on barges, but low water delayed the boatmen. When I arrived to see how things were progressing I found the tractors and grain drills parked motionless at the edge of the basin. The drivers anxiously paced their quarters.

To wait for boats that never came was intolerable. I finally decided to set out for Yungnien. There another farm, organized a few months before, had been flooded out and rumor had it that twenty drums of gasoline had been stowed away. The trip turned into a comedy of errors. I set out with Chiao Shih-ju in a new Dodge truck. The *Ta chi*, as it is called in China, is the favorite truck. Chiao was happy to be in one. But even a *Ta chi* can't steer itself. We lost our way. When we finally got to Yungnien we found that the farm manager was not the least interested in helping out Chiheng. With his own land under water he had problems enough. We tried to persuade him that we would soon return the borrowed gasoline. He was not moved.

In the city of Hantan we had a stroke of luck. We located vast stocks of UNRRA oil in the supply department of the government transportation company. We bought the oil, gave some of it to Yungnien's sour chief, and got in return twelve drums of gasoline. With a trailer loaded with oil and a truck full of fuel we started back to South Ridge.

Half-way home the truck slowed down and stalled. Try as we would we could not get it to start up again. It was October 1, the day of the founding of the People's Republic of China. In every city and village and town of North China joyous celebrations were under way welcoming the dawn of a new era. In Peking a seventy gun salute ushered in Chairman Mao's announcement: "The Central People's Government Council of the People's Republic of China took office today in this capital."

A vast parade of soldiers, workers, students, and ordinary citizens from town and country surged across the square in front of the Gate of Heavenly Peace. Two hunded thousand people assembled to watch the parade. Dancing and rejoicing continued until dawn.

As we struggled with the recalcitrant motor, Chiao and I could hear firecrackers going off in the villages that lay around us on the plain. We spent the night in a cotton field. The next day a message sent through to Chihsien brought Lao Hei on a tractor. He towed us home with a flourish as peasants strang along the road shook their heads at the sight of the little Ford tractor lugging behind it such a huge truck and heavily loaded trailer.

"That's truly a *tao mei chi ch'i* [a hard luck machine]," said one old man as we roared through a narrow village alley.

The final irony was yet to come. When we got back to South Ridge the barges had arrived. As we rolled across the last flat we saw the grain drills at work as if nothing had ever delayed them.

When I thought of all these things I burst out laughing. Now that they were over they seemed funny. Just then militiaman Kuo jumped up on the trailer and sat down beside me.

"We won't get finished today," he said. "Shall we keep going all night?"

"How are the lights?"

"There ought to be enough good lights to equip two tractors. I'm only afraid they won't be able to see the mark." He meant the furrow cut in the ground by a steel disc attached to the drill. It marked a line for the tractor wheel to follow.

"I think they'll see the mark all right," I said. "Let's try it."

Kuo passed the word around that work would continue until the job was done. The drivers were in favor of it. They gassed up their machines, checked the oil and water, and then knocked off for half an hour to eat supper. By the time night fell they were at work again.

The night was clear and brilliant. So many stars came out that it looked as if a bag of millet had been spilled across the sky. Whether it was the moisture in the air, or the coolness of the season, or the superior quality of the oil that we had unearthed in Hantan I do not know: whatever the reason, the tractors purred on hour after hour without a halt or break, as if trying to demonstrate at last their dependability.

Lao Kuo and I sat on the trailer and wrapped ourselves in cotton quilts to keep warm. When the seed boxes on the drills were empty we helped load them full again. Other than that there was nothing to do. We fell to talking about the days when Kuo fought in the Taihang militia.

He taught me a song that went like this:

> Taihang mountains
> High, oh high!
> A hundred times ten thousand men
> Take up the cry.
> Young men of the soil
> Fear not shell or knife
> Each shot we fire
> Takes an enemy life.

"Were there really a hundred times ten thousand men fighting in the Taihang?" I asked.

"Of course, maybe even more," said Kuo, "and that's not counting the regulars. Our mountains are perfect for guerrilla warfare. Every village had twenty or thirty organized men.

"In our county there is a cave, the famous Licheng cave. No one knows how deep into the mountain it goes, for no one has ever been to the end. A river of clear water flows out from its mouth. We used to hide there when the *kuei tzu* [Japanese devils] came.

"It was right near that cave that we won our first battle. When the burn-all, loot-all, kill-all campaign began, everyone was frightened. Many people thought the *kuei tzu* were invincible. We kept a careful watch and as soon as troops were sighted our people hid what had not already been buried and headed for the mountains.

"But we young men got tired of that. We wanted to strike back. The older men said no, resistance would only bring reprisals, but we organized a militia group anyway. I was fifteen years old at the time. They elected me captain because I was the toughest boy in the village, even though I was small. We didn't have any weapons, only two old rabbit guns and a rusty rifle. We concentrated on stone mines. Did you ever see a stone mine?"

"No," I said, "but I have heard of them."

"They are easy to make. Take a rock, hollow it out, fill it with powder, and plug it up tight. Of course you have to connect a fuse. Then you pick out a place where the enemy is likely to come. The mines are buried carefully in the path. Someone has to hide out near by to shoot the blast.

"The first time we used them we picked a place by the river. The path, after going through a narrow gorge, widened in a little grove of trees. We set our mines cleverly all around the grove. The *kuei tzu* came and, just as we had hoped, decided to rest in the shade of the trees. They took off their packs and sat down, some of them right on top of the mines.

"There was a terrific explosion. Smoke billowed toward the sky as if the whole grove had suddenly caught fire. Some of the *kuei tzu* screamed with pain while others grabbed their guns and fired wildly at the hills on all sides. But they didn't dare comb the slopes. Too many were already dead. Of the original thirty at least twelve were done for. Another six or eight were wounded. Those who could still walk picked up the injured, left the dead lying where they fell, and fled.

"From that day on no *kuei tzu* patrols dared come near our village. We won great prestige with the people. Our group became famous in far away countries. We got very bold and roamed the mountains looking for a chance to hit again.

"Once I was in a village, sleeping on the roof of a shed, when the enemy surrounded the whole place. I jumped down into the courtyard and hid in a stack of beans. They searched every house but didn't find me. That was the advantage of being small.

"Another time we got caught in the mountains. A large *kuei tzu* patrol cornered five of us in a cave high on the slope of a ravine, but they couldn't get at us. They fired at us but they couldn't hit us. They couldn't get any nearer either because to go forward they had to cross a narrow ledge where we could pick them off. We stayed there all day without food or water. At nightfall they left. They never liked to be out after dark."

Kuo's story broke off as Liu Po-ying came looking for me.

"I want you to check the water pump," he said. "The fan is wobbling."

The fan was wobbling, but I thought it would last the night, so I told him to go ahead. Already the drills had sown their way halfway across the field.

Chapter 17

Militiaman Kuo Passes the *Gate*

WHEN I GOT BACK TO THE TRAILER Lao Kuo was already thinking about something else.

"I want to ask you, Lao Han," he said, "can you imagine me as an arrogant, short-tempered person?"

"No," I said, "certainly not." The very idea was absurd. Lao Kuo was always calm and quiet. Even when everyone else got upset he sat quietly and sized up the situation. He had a way of talking to people that helped them to relax. He could bring up the sharpest criticism in such a way that the person criticized was able to see sense in it. That was because he looked at everything rationally, with a minimum of emotion. And yet Kuo was not a cold person. He was full of drive and fire, wit and song. He loved people and understood them well. I always found it hard to believe that he had never been to school, never read a novel or listened to a radio. He seemed so sophisticated.

"Well, I used to be very proud. I lost my temper all the time," Lao Kuo said. "The victories we won in the militia went to my head. I thought I was a hero. I carried two guns and swaggered through the villages at the head of thirty men like a baron or a duke. When the people didn't do as I told them I got angry and cursed them. I had a girl in every village, sometimes more than one. I was crazy about women. When I wanted to visit a girl I posted militiaman outside the house to warn me if anyone came. My wife was so distraught that she tried to kill herself.

"The people had a name for me: 'The Little Emperor.' And it was true. I forgot my origin. I thought I was better than other people. And all because I had killed a few Japanese."

"What happened?" I asked. "You've certainly changed since then."

"Yes," said Lao Kuo, slowly. "I changed a great deal. I had to learn to control myself and to listen to others."

"But how did you do that? It's not very easy to change one's character."

"I had to. I had to pass the *gate*."

The *gate* was a form of examination by the people, by the peasant masses, of all leadership, all cadres. In 1948 the Communist Party opened its regular self-and-mutual criticism meetings to the public. Throughout North China the Party called for delegates elected by the peasant unions to attend Party meetings, offer criticism, and supervise the work—past, present, and future—of all Communists. Non-Communist cadres were also invited to these meetings and were criticized in the same manner. Peasant delegates had the right to repudiate any cadre and deny him or her the right to hold office again.

This whole movement was part of a general review of the progress of land reform after three years of turbulent uprising and struggle against the ruling gentry in the villages. It was an expanded form of Party rectification made possible by the fact that the Liberated Areas governments were at last relatively secure from counterattack and could with some safety reveal the names of all leading cadres.*

"During the Party rehabilitation movement," Kuo said, "I had to stand up before the delegates of the village and make a self-examination. I defied them at first. Hadn't I risked my life dozens of times for them while they hid in the safety of the mountains? Who were they to criticize me? And why should they hate me so? But they shouted: 'What is your thought? What makes you think you can climb on the backs of the people?'

"For almost a week I didn't sleep. I hardly ate a thing. My mind was in a turmoil. It was clear that no one would accept me if I didn't change. The people would no longer have anything to do

* For a detailed account of the *gate* in a North China village in 1948 see *Fanshen: A Documentary of Revolution in a Chinese Village*, Parts IV and V.

with me. Nor would the Party. If I did not bow my head before the masses I would have to return home and farm.

"I went before the *gate* again and apologized for some of the worst things I had done. But still they were not satisfied. For I was still proud and I resented their criticism. In my heart I did not accept it. Either they were all wrong or there was something wrong with me. I had to think it through. I saw that I had mounted the horse. From a servant of the people I had turned into a tyrant. If I went on that way I would be no better than the landlords' bullies whom we all hated. I realized, too, that I could not have fought the Japanese alone. How could I have done anything without the people, those who stood guard, those who raised the food we ate, those who hollowed out the mines? Our victories were the victories of the whole village.

"I felt ashamed. When I went before the *gate* for the last time, I spoke about everything that was in my heart and the people forgave me. From that time on I have been a different person. It was the biggest lesson of my life."

"Lao Kuo," I said, "do you think everyone can change and grow?"

"With the help of the people almost everyone, yes. It is like the hairs on my head. Among the countless black hairs only one or two are white. Just so, among countless people there are only a few who cannot, who will not listen and reform themselves in order to carry the revolution forward."

After loading another grain drill he told me an almost incredible story to illustrate the point.

"In our mountains there was once a famous outlaw whom we tried to reform. He was twenty or so when I first knew him. He could outrun a horse. He could leap farther than a mountain lion. He could lift boulders that no one else could even move. Many times when he got into trouble with the security forces he escaped by jumping from high places that no one else even dared to climb down. He leaped ravines as if they were puddles.

"He could do the work of five men, but he didn't like to work. And he didn't have to. He became a bandit. A few friends followed him. For years they lived an easy life robbing wealthy landlords and merchants. They were bold beyond belief. Once Yang, for that was his name, went into a county town and robbed the magistrate in broad daylight. No one dared cross him, for he was quick to anger and had killed more than once.

"When the Japanese came, we tried to get him to join us. He was a wonderful guerrilla fighter, fast and tireless. But he didn't like the discipline. He went along with us for a while, but he suddenly took off on his own, fought the Japanese in the morning and robbed the wealthy in the afternoon. The robberies gave our guerrillas a bad name.

"We sent a squad out to capture him, but even with the co-operation of the people they couldn't catch him. His men had deserted him, and he was still more than a match for us. The only way we finally caught him was to persuade a sweetheart of his to entice him to a rendezvous. He trusted her and came. It took six men to tie him up. We left three men to watch him. He broke out of his ropes, overpowered the guards, and leaped from the roof of the building into the street. His guards said he hurt himself when he landed but he still ran faster than they could.

"After that he lived wild in the mountains. We gave him up as hopeless. Then one day he killed one of our best fighters, a man with whom he had quarreled over a woman. That made us angry. Robbing we could endure, and loose living, but when he started to kill our own people it was too much. The word went out: Bring him in dead or alive. We tried to capture him alive but he vowed he would not be taken. In the end he was shot while running down a ravine. When the militia reached him he was dead.

"We thought we had rid the country of a menace, but when Commander-in-Chief Chu Teh heard about it he was very upset. He said we were very backward. He said that we had not made enough of an effort to win Yang over. He said that with patience and more effort Yang could have changed his outlook and become a great fighter for the people. Such people as he were rare enough, one in many many millions. It was a great crime that we had committed.

"I think Commander-in-Chief Chu was right. The power of the people is very great. Properly organized we could have won Yang to our cause. We could have reformed him. He sprang from the same origin and suffered the same oppression as the rest of us. Instead, we drove him to the cliffs and the caves and finally shot him like a wild animal. That was the result of our impatience."

The story of the bandit Yang was indeed a strange one. As Kuo told it I felt the loss of the man as if I had known him. For a long time we sat in silence watching the stars. A thin moon rose in the east and began to light up the basin. The tractor lights paled by comparison. The character of Kuo himself set me thinking. Every time I talked to him I learned much that I had never known before. There was something about his outlook on life that was more mature, more steadfast, and more enlightened than my own. At twenty-two life had taught him more than books, colleges, and world travel had taught me at thirty.

Liu Po-ying's tractor pulled up. The grain drill was empty and we helped him fill it. The seed on the trailer was almost gone, but so was the unplanted land. A few more sweeps back and forth would finish the field. In another half hour the first mechanized crop in North China would be in the ground.

"They are almost finished, Lao Kuo," I said. "How do you think it will turn out?"

"I'm worried, Lao Han," said Kuo, peering out across the flat. "There is too much salt everywhere, and not enough fertilizer."

"What worries me," I said, "is how we are going to harvest all this. How are we going to cut twelve thousand *mou* with sickles?"

"If it comes up we'll harvest it, don't worry about that," Kuo said.

But I worried about it just the same.

Chapter 18

Double Bridge

ALL THROUGH THE SUMMER AND FALL of 1949 the roads leading
out of Peking rumbled and shook with a never-ending stream of
trucks, carts, and handwagons hauling muck and trash from the
city. Thousands of soldiers stood waist deep in slime scooping
the silt from lakes and moats that had not been dredged in the
memory of the oldest inhabitants. They transformed stagnant
pools filled with swamp grass and mosquito larvae into clear
lakes that fish could live in and boats sail on. Municipal workers
in blue tackled sewers that had been plugged for a hundred years,
while people cleaned out abandoned courtyards where mountains
of trash made stinking citadels for rats, mice, and lean, scabby
dogs. Alleys that had been blocked with refuse since the time
of the Boxer Rebellion were made into broad, level walks. All
the rich material removed from the city was pressed into service
as fertilizer for the next year's crop on the plain surrounding
Peking. The director of the Peking Municipal Planning Bureau
later said that 600,000 tons of garbage were removed.*

Trucks and carts dumped their reeking loads at the edge of
the highway a few miles out of town and the peasants came from
near and far to haul it away. Those who owned carts came with
carts, those who owned wheelbarrows came with wheelbarrows,
those who only had carrying poles brought their poles and
baskets. Women and children, old men and young boys, came
out to gather up the foul remains of decades and centuries of
Peking corruption and cart it off to the land to be transformed
into golden grain and green vegetables.

---

* See Robert Trumbull, *This Is Communist China* (New York: McKay,
1968), p. 200.

Everyone was happy to see the filth go. As the heavily laden vehicles rolled toward the city gates neighbors called to one another: "There goes the last of the Kuomintang filth."

It was between rows of carts that Sun Hsiao-kung, the interpreter from the tractor class at South Ridge, and I drove out to Double Bridge, ten miles beyond Peking's East Gate (the famous *chiao yang* or "facing the sun" gate) to find a site for a second winter training class. What we discovered was the place that was to replace South Ridge as the center of tractor work in China.

To return to South Ridge in Central Hopei in the winter of 1949-1950 made no sense. At least ten new farms had already been established by South Ridge students during the summer and Peking was central to all of them. All the departments of the new national government had migrated to sites in and around Peking and all nationwide projects centered there. Our training class in 1950 was not sponsored by the government of North China but by the State Farm Management Bureau of the Ministry of Agriculture of the People's Republic of China. As a consequence, although we would still return many times to South Ridge, and although our first major crop lay in the ground there, we decided to train new drivers and give advanced courses to old drivers at a site somewhere near Peking. Rumor had it that there were some run-down buildings on a city-owned farm at Double Bridge. We found that the rumors were not completely unfounded. There were some old buildings, but "run down" was much too mild an adjective. Most of them had long since collapsed and the whole place had been looted.

What we found was a farm that had once been a barracks for the troops that guarded the radio station and central telephone exchange of the Japanese armies in North China. To make the installation safe the Japanese had quartered several companies of soldiers there and had confiscated hundreds of acres of farm land so that the people could be forced to keep their distance. The commander had set his men to work raising vegetables and hogs for the officers' mess.

The Japanese erected their barracks in a walled compound doubly protected by a moat and a high voltage electric fence. Even this was not considered safe enough for the higher brass.

In the northeast corner of the compound a second square had been walled off and fortified at the corners with concrete blockhouses. Here the Japanese officers evidently felt safe, not only from the Chinese but from their own men as well.

The officers must have brought their families with them, for their dwellings were divided into apartments with all the comforts of home. Hsiao-kung snorted when he saw them. Each unit had a kitchen and next to it a privy, both under the same roof. "The Japanese don't know the first thing about sanitation," he said. "They always have the cooking stove and the privy side by side."

The roofs of the buildings had long since collapsed. Rubble and trash littered the ground. The new caretaker lived under one of the few roofs that remained. One large brick warehouse was still intact and, to the west, not far from the compound wall, was the huge, barnlike hulk of the former telephone exchange. Tiles had fallen out of its roof leaving gaping holes and all the windows had been smashed, but the walls looked sturdy and the doors large—large enough for a ten-ton tractor to go through.

We decided that these big buildings would make excellent classrooms. Roofs could be replaced on the ruined barracks to convert them into dormitories. The land would be a convenient practice ground for tractors.

We returned to town full of enthusiasm for Double Bridge. There, only a few minutes outside Peking's walls, we had found the answer to our needs.

Work began on the buildings at once. Hsiao-kung took charge of this. The first things he ordered built were some large and well-appointed privies, far removed from any other buildings. Director Li, who was appointed to head up the new training class as he had the old, made out long lists of the lumber needed for beds, the cooking pots and utensils needed for the kitchen, and the office supplies, ink, and paper needed for the school. While he was still interviewing accountants and office boys, letters went out to all the mechanized state farms in the country to send in some, or all, of their drivers for winter training as soon as the ground froze in December. And, since the farms had few good tools, measuring instruments, or men with the know-how to supervise repair work, the mechanics on every farm were also instructed to come to Double Bridge, and to bring their tractor motors with them.

Chiao Shih-ju, the man who had gone with me to Hantan in search of oil, was put in charge of repair work. Stored corn and wheat belonging to the farm were moved out of the one good warehouse so the building could be converted into a shop. Lathes and drill presses were bolted to the floor. Work benches were made and covered with tin. Almost overnight the place took on the air of a professional garage.

Chiao, in a creative streak, converted a big oil drum into a stove that heated the building, provided boiled water for drinking, steam for cleaning, and distilled water for batteries, all at the same time. He named this the *wan neng lu* or "ten-thousand abilities stove."

Even before the new roofs had had time to set on the buildings the drivers and mechanics began to arrive from the countryside. There were many joyous reunions of South Ridge people who had not seen each other since they had plowed the Thousand Ching Basin together that spring, for while one contingent had remained at South Ridge, the majority of the trainees had set off in midsummer to establish new farms.

The drivers from Lutai told a discouraging story of soil and sod so heavy that the tractors could not plow it up, and field levees so high that the tractors could not cross them. The men from Chahar, who looked like trappers from the wild west with their fox-fur hats and heavy wool-lined overcoats, were elated with their success. With ten Allis Chalmers tractors equipped with disc plows they had plowed 30,000 *mou* in the few weeks left to them before the ground froze that fall. In the bitter cold they had driven in twelve hour shifts, kept the tractors going night and day, and turned over all the virgin land on the farm in preparation for sowing a spring crop.

Those who had remained behind at South Ridge told of putting in over 10,000 *mou* of winter wheat and then finishing off another 10,000 *mou* of fresh fall plowing. They said the wheat was coming up well and a good crop was anticipated. Yungnien drivers had also planted a large area of wheat, but complained that the Fords didn't have the power to plow out the heavy roots in the rich basin around the county town. The same complaint came from Poai, way down by the Yellow River. On that farm, they said, the grass roots were as big as a child's arm and the ground was full of humps and hollows. It was almost impossible to do anything with the Fords.

Many new faces appeared. Technicians and drivers came from Central and East China, from Kuangpei, the new farm which was being set up on the wastelands near the mouth of the Yellow River, from Tunghsin in north Kiangsu, from the old UNRRA/AMOMO warehouses on Fire Island in Shanghai, and from farms and farm tool shops in Chekiang and Kiangsi. A big contingent also arrived from Honan, the old Yellow River flood area where the main UNRRA tractor effort had been concentrated. Soon we had more than two hundred drivers and mechanics.

The first few weeks were spent reviewing the past year's operations. It was a time to sit down and think over all that had happened since the tractor class ended in March, to discuss what had been done well and what had been done poorly, and to criticize those responsible for waste and mistakes. Since most of the farm managers and the brigade leaders were also present, it was possible to examine operations from top to bottom.

The South Ridge group discussed again the disastrous mistake of using a cheap engine oil that had resulted in the simultaneous breakdown of the bearings in more than ten tractors. The crews from Nanyuan, the South Airfield farm, had a lot of criticisms to make of their manager, Jung, who ran things in a high-handed way and made decisions without consulting anybody. Some of those decisions had resulted in big losses for the farm, such as the rotting of almost 100 tons of sweet potatoes because of inadequate storage.

Supply parts, living conditions, work clothing, management, planning, and a thousand and one other questions came up. Everyone had a chance to air his views and gripes and to make suggestions for improvement. A large number of recommendations were turned in to the various farms and to the Farm Management Bureau.

In the middle of this review I went off to Harbin in Manchuria to bring back the Soviet tractors, plows, and other implements that had been promised the national bureau by the State Farm Bureau of the Northeast. (In the Northeast, farms had been operating with Soviet tractors for more than two years.) The Northeastern bureau had more equipment than it needed for the time being and had been persuaded to lend a few items to the central government until new machines could come from Russia to replace them. Most urgent for our work were machines with which to harvest, or at least to thresh, the wheat crop at South Ridge.

Chapter 19

Soybeans for Steel

CHOU PU, A MECHANIC, Li Kun, an administrative cadre, and I arrived in Harbin on a December morning. We stepped off the train into the most biting cold that I have ever felt. Our feet went numb and our lungs winced at the intake of sharp air.

Snow lay all around and squeaked underfoot. We could think only of one thing, to get to the state farm headquarters as soon as possible. We hired a droshky pulled by a stumbling old horse and driven by a man in a thin padded coat. His feet were encased in huge felt boots but that was the only part of him that appeared to be well clad. How he survived the cold and rode around on that high seat all day was hard to imagine.

The bureau headquarters were on Asheke Street, in what must once have been the Russian section of town. In the bright morning sunlight we seemed to have come to a foreign city. Hills, wide streets, great plaster-covered houses and halls painted in pinks, blues, and greens stretched out on every side. After the tight alleys of Peking and the crowded streets of Tientsin and Mukden, Harbin looked like a park. Perhaps the snow had something to do with it. Under the blanket of crisp white crystals everything looked immaculately clean. The trees along the sidewalks added to the pleasing appearance of the sector.

The state farm office put us up in a typical Russian-style Harbin house. It was one story high and solidly constructed, with walls at least two feet thick. It was divided into sections of four rooms each with a round brick stove where the four rooms met so that each room shared a quarter of it. The quarter of the stove that was in our room radiated heat in waves, which made the indoor temperature as hot as the outdoor temperature was cold. Fortunately, a small panel in the double window provided enough air to prevent us from suffocating while we slept.

A Comrade Ku and his wife were in charge of the whole Northeastern state farm program. We liked them immediately. Ku was a tall, robust, plain-spoken man who never seemed to be in a rush. The desk in his enormous office had nothing on it. He apparently had plenty of time to talk. I got the impression that he was a good administrator who helped to decide the main direction of the work and turned the details over to the rest of the staff.

Ku's wife was small, handsome, and lively. She wore her hair in long braids wound around her head. As administrative secretary of the bureau she seemed to have a grasp of the details of the work which surpassed that of her husband. Between them they made a very competent pair. They had walked all the way to Harbin from Yenan after the Japanese surrender in 1945. (This was part of the race to reach Manchuria before the Kuomintang, with American air support, could fly in enough officials and troops to take over.) Even though neither of them knew anything about machinery or farming they had been chosen to head up the agricultural mechanization program because of their ability to lead others.

After seeing them in action I came to the conclusion that technical knowledge was the least of the requirements for heading up any large enterprise, whether it be agricultural or industrial. The important thing was to know how to get people together, allow them to thrash out their problems, and guide them in choosing a correct solution. This technique the Ku's had mastered.

Their first regional production meeting had just concluded. The rooms buzzed with the activity that would put the plans for 1950 in motion. Compared with North China, the Harbin bureau was very well organized. It boasted a supply department and a technical department. The technical department was staffed with agronomists, mechanical engineers, draftsmen, and translators. The chief translator was a German who had been the main dealer in Lanz tractors in Manchuria for more than twenty years. He translated the Soviet tractor and machinery manuals into English and then a Shanghai-trained engineer retranslated them into Chinese. They had worked hard for a number of months and already had more than ten manuals, complete with illustrations, in print.

The German was a typical Hanoverian businessman. He obviously did not think the Chinese were capable of setting up and operating large state farms. He regarded the Soviet tractors and equipment as a bunch of junk. He described a week he had spent on a state farm the previous year. He said it was a mad rush, night and day. He no sooner settled down for a little rest when a new breakdown brought him to his feet again. And so he went rushing from one field to another trying to keep the tractors running. All the troubles he blamed on the Russian equipment. "You can't rely on any of it. I never saw anything like it," he said. "Ordinarily a carburetor, say, will develop certain troubles that can be predicted. So will a magneto. But these Soviet jobs, they're completely unpredictable. You never can tell what might go wrong. So it is almost impossible to teach the drivers how to repair them. You can't lay down any rules at all." Of course the German-made machines were a different story, according to him. They never broke down. He intended to go back home to the Lanz factory in West Germany. Already they were building a big trade with South America, Africa, and southern Asia. Business was booming. He couldn't wait to get back.

The Shanghai engineer had studied for many years in England. He was not happy about his work. He was an expert on internal combustion engines and didn't know very much about tractors and farm machinery. He had worked in an English automobile factory and regarded duty in far off Manchuria as a sort of exile. He found the climate much too rigorous. Also, like all engineers I met before or afterward, he longed to be in a factory, working in production. I don't think I ever met an engineer in China who was content to help others use machines more efficiently and keep them in good repair. They all wanted to be industrial designers turning out machinery in large-scale industry. Nevertheless, this engineer had led his department in trying to solve some of the problems met with on the farms. They were making drawings for a wide boom which a tractor could use to pull several implements. They had also designed a type of plowshare more suitable for turning over heavy prairie sod than the shares that came with the Soviet plows. In addition, as already mentioned, they had translated and printed up numerous instruction manuals. Of these last we ordered large quantities so that when the Soviet machinery came to North China every driver could have a copy.

Almost the first thing we did was to visit the big supply depot out at the edge of town behind the power plant. Here were warehouses and lots filled with farm equipment and tractors, huge combines and horse-drawn reaping machines, and spare parts for them all. Hundreds of machines were lined up in rows in the snow, some of them in pieces and others fully assembled. A barbed wire fence surrounded the lot and a soldier at the gate let no one in without a special pass issued at the supply base office. After the scrimping and figuring we had been through in the past year trying to make the best use of every machine and wondering how we would ever keep them running without any parts, this depot was incredible. Here was everything a farm could need. Like children at Christmas time we could walk in and take our pick.

And all of it was said to be very cheap. It had been acquired from the Soviet Union in return for Manchurian soybeans at very reasonable rates. I was told that a carload of beans could be swapped for a Stalin 80 tractor. Since the warehouses of Manchuria were bulging with beans, this was a very attractive proposition and the State Farm Bureau had taken advantage of the situation to buy what looked like enough machinery for several million acres.*

Ku was willing to give us six new and four used STZ 54 horsepower kerosene-burning crawlers from the Stalingrad tractor works. We looked them over with intense interest. What struck me first of all was the extreme simplicity of the engine, the drive mechanism, and the track. A glance was enough to reassure me that here was something we could handle. There was nothing new or unknown in principle, nothing that our students had not

---

* Chinese experience with Soviet prices on industrial goods in later years contradicts this. In the fifties Soviet-made machines carried exorbitant price tags and Soviet trading organizations demanded big mark-ups on machinery produced in Eastern Europe.

already studied, in the construction of the engine. It was unusual only because of its great size and weight. I immediately felt better. The idea of bringing home a shipment of Soviet tractors that we could not handle had been worrying me for some weeks. We took a closer look. What was Russian machinery like? How did it compare with that from the United States? All the world was interested in that question. Well, it certainly was not beautiful to look at. The tractors were built square and straight, the castings were rough and had a lot of corners and sharp edges. The tracks were cast steel plates linked together with pins. A great crank almost two feet long stuck out from the frame in front and hung to one side suspended on a bent wire. Instead of fins to control the cooling there was a piece of canvas on an iron rod that could be raised or lowered at will in front of the radiator.

No, the tractors were not beautiful, but they looked very sturdy. They were built for use, not for display at international fairs. They had a straightforward, stripped-for-action appearance. No fancy grille, no streamlined cowl, no extra gadgets. Just the essentials. They were painted a dull gray.

I thought it would be a formidable job to start one up, especially at twenty below zero. It did not seem possible that one man could turn that huge crank and bring the cold machine to life even in summertime. But in that respect appearances were misleading. One of the mechanics at the base got a little can of gasoline, poured a few drops into two especially constructed ducts which led the gasoline into the intake manifold, and gave the crank a wide swift swing. Our ears were greeted with a deafening roar. Flames and smoke belched from the four-inch exhaust pipe. The motor was turning over. It soon settled to an even rat-tat-tat like a machine gun going off very near at hand. The mechanic ran for water to fill the radiator.

We were to become acustomed to that sound. We were to learn to tell all sorts of things about the engine itself and the man who was driving it from the rhythm, the tone, and the free or restricted outpouring of that staccato exhaust. But when we heard it for the first time it only seemed terribly loud. How was one ever to listen for engine knocks, or transmission noises, with that devil's concert going on up front? The vibration made the bolts and plates of the cab sing and hum. As the radiator filled, the sharp shaking of the whole tractor frame made the water come dancing out at the top in little geysers.

But it proved to be very maneuverable. It surprised us with its speed. In high gear it pranced and clanged along at better than six miles an hour. A pull on the steering clutch levers sent it into sharp zigzags. We were told that with a load on the drawbar it turned even more easily.

The one thing we were warned about was to be sure to have the engine warmed up properly before switching to kerosene. A temperature of around 200 degrees was needed before the kerosene could vaporize and burn well in the cylinders. They had had a lot of trouble because inexperienced drivers turned on the kerosene valve before the engine had warmed up; it then skipped and popped, piled up carbon in the combustion chamber, and leaked fuel past the rings into the crankcase. It didn't take long to so thin out the oil that a change was necessary. "Run her hot," was the advice. "Keep the cloth over the radiator until you see steam rising from the cap. Then change to kerosene and keep her that way." Thus we picked up our first tips on the Soviet machine.

We picked out our ten STZ's and then went on to pick plows, disc harrows, spike harrows, grain drills, cultivators, and winnowing machines. On close examination all the equipment showed the same characteristics as the tractors: rough, rugged, utilitarian construction, simple design, no frills or extras. We were very pleased and could hardly wait to see one of those big multiple gang plows turning five black furrows ten inches deep as it went down a field. The Northeast bureau had agreed in advance to give us all this equipment so there was no trouble about it. Anyway, they had enough there in the lot for quite a few more farms. No one would miss ten sets, or even twenty. But there was one problem that we couldn't get off our minds—the wheat at South Ridge, the 10,000 *mou* we had finally sown with such great effort. How was it ever to be reaped, and once reaped, how th shed? We did not have the nerve to ask for a combine. They had ten or twelve just in from across the Urals, but they had tens of thousands of *mou* of wheat planned for the spring and the combines would be hard pressed to handle the crop. We finally decided to ask for one of the big threshing machines that stood way off on the side of the lot: we thought that with all those new combines no one in Manchuria would be interested in threshing machines.

We proved to be wrong. After an hour or two of urgent pleading Ku agreed to give us one, but when the request went downstairs the man in charge of the supply department refused to honor it. Instead he rushed up to see Ku and was closeted with him for more than half an hour. When Ku called us in again he apologized and said that he would like to give it to us but the plan would not allow it. Every bit of threshing capacity in the Northeast would be needed to handle their crop.

This was a terrible blow to us. Comrade Li and Chou Pu were willing to drop the matter, but they had not been at South Ridge. They had not made a five-day trip to Hantan in search of gasoline, or stayed up night and day putting in a crop of wheat. I said we had to try again, otherwise the whole ten thousand *mou* at South Ridge would have to be cut and threshed by hand. And what if we couldn't find all that manpower?

So we went at it again. Li talked with Ku while I talked to his wife. I described the wheat in Hopei in glowing terms. I told her all the difficulties that we had gone through to start it off right. I asked her how it would look if our first harvest, the first mechanized wheat south of the Great Wall, were to be cut by hand and rolled out on the threshing floor with stone rollers.

I listed for her every piece of machinery that we had so that she could understand clearly the pitiful shape we were in on that score. I asked her if it were not *pen wei* to leave us without any threshing equipment while the Northeasterners used brand new combines. This was a telling argument. *Pen wei* meant local sectarianism, a great effort to keep one's own house in order but complete apathy when anyone else's problems came up. Cadres from the old Liberated Areas, where the forces had been isolated from one another and each had fought desperately for its own life, knew its evils well.

In the end she gave in. She talked with her husband. He ordered the supply man to release the threshing machine. The solidarity of North and South prevailed.

Our life in Harbin was not all hard work and special pleading. New Year's Day 1950 (new calendar) came and went while we were there and on New Year's Eve the whole city of Harbin celebrated. We were invited by some Russian friends whom I had worked with in North China to the Teacher's Union ball. We danced until dawn in a gaily decorated recreation hall where revolutionary young people, full of vigor and enthusiasm, never seemed to tire. Only the few Russian girls came in dresses; the Chinese girls, already teachers though still in their late teens, wore black serge trousers and Lenin jackets. Their hair, bobbed to fall just short of the neckline and cut straight across the forehead in a bang, framed each radiant face in a square of jet black. The thirty-below cold outside and the eighty-degree heat inside the hall combined to bring an extraordinary flush of healthy red to their faces. Flashing black eyes completed the lively picture.

For the first time in my life I felt old. By midnight I was tired, but the young people danced on until dawn and when I tried to sit out a dance they would not allow it. As the only American in the whole Northeast I was a celebrity and every girl in the Teacher's Union demanded a dance with me.

A few days later Chou Pu and I visited a state farm several hundred miles north of Harbin, halfway to the Amur River. It was situated in the heart of an extraordinary rolling plain, reminiscent of North Dakota, with virgin black soil three to five feet deep just waiting to be plowed. They showed us samples of this soil. If you put a few ounces in a jar of water and shook it, the water remained clear while the soil particles remained black, so tightly bound were the molecules. The only drawback to large-scale farming here were the numerous bogs and swamps in the hollows between the rolling uplands.

We saw neither swamp nor soil because everything was frozen solid and covered with two feet of snow, but we did learn how they had organized their farm, how they maintained their machinery, and what they planned to do with the land in the next few years. Their plans were bold and exciting, for here there was no problem of fertility, no alkalinity, no chronic flooding, no history of drought—only millions of acres of chernozem waiting to be tilled. That night the temperature went down to forty degrees below zero but no one even talked about it. The harsh climate seemed to make people healthy. Of all the people I saw in China, the Northeasterners seemed to be the most rugged and the most vigorous.

Chapter 20

Wherever the People Need Us

"First soviet tractors arrive in capital"—so read the headlines in the *People's Daily* on that day in early January 1950 when the ten carloads of tractors and farm machinery from the Northeast finally rolled into the Double Bridge station. The sun shone brightly. The air was clear and cold. The whole student body rode out to the siding to speed the unloading, for the stationmaster insisted on clearing the cars before noon. All hands fell to with a will. The heavy iron wheels and beams and boxes were lifted bodily from the gondolas while mechanics drove the tractors off. As each one reached solid ground a great cheer went up.

Everyone was delighted with that sharp, staccato exhaust, with the tremendous roar of the engines, with the flashing, clanking tracks, and the speed and snap of the tractors on the road. Right then and there the prestige of American tractors and of American industry itself shrank several sizes. Here at last was a tool suited to China's wastelands. Here were plows of giant size, grain drills that truly spanned a field. From that moment on, not one of the hundreds of students or staff members at Double Bridge was ever interested in a Ford tractor again.

Lin Chi was so pleased he lectured everyone. "You see, that's the difference between the U.S.S.R. and the U.S.A. In the U.S.A., they make those little Fords, useless little beetles that can't even plow more than six inches deep. But in the U.S.S.R., look what they make—54-horsepower crawlers. Now there's a tractor that'll really do a job. It'll pull five plows and plow ten inches down. That's the difference between socialism and capitalism for you."

I told Lin that America made big tractors too, in fact even bigger ones than any we had yet seen, but that didn't make much of an impression on anybody because there were no concrete examples sitting right before their eyes. What machines came from America? The little Fords. What machines came from the U.S.S.R.? The huge tracklayers. And I think Lin Chi was right, for the Soviet 54-horsepower crawler represented Soviet agriculture of that decade just as surely as the 20-horsepower Ford represented contemporary American farming. The fact that America also made large crawlers was really beside the point, for they were not widely used on private American farms because the fields were too small. Compared with Soviet collectives, American farms were miniatures.

A group of photographers and reporters drove out from the city to record the historic unloading. They snapped pictures right and left, interviewed Director Li, talked with those who had brought the tractors from Harbin, and even lent a hand unloading parts. The crowning shot on their film showed all ten tractors in a row clanking down the road toward the school gate while peasants, students, teachers, and staff workers lined the banks to welcome them.

The arrival of the tractors meant that the work of the winter could begin at last, and there was no time to lose. We had only two months to learn how to make these monsters work and keep them running. No one had ever seen a kerosene-burning tractor before. Who would do the teaching?

Director Li organized the best qualified technicians and mechanics from each farm into a "teaching group." Among them were:

—Hsiang Tze-jung from South Airfield. People teasingly called him the Japanese because his round face, his horn-rimmed glasses, his low forehead, and profuse growth of chin whiskers made him look like a man from Nippon. He had studied mechanical engineering in Tientsin, worked for a number of years in auto repair shops, and then in 1948 joined a Nationalist-run plowing project in the Pohai Gulf area.

—Ma Ching-po, an old classmate of Hsiang's who was working in a Nationalist army truck repair depot near Hankow when the Liberation Army swept through. I was in the bureau office the day he came in looking for a job—young, clean, eager. As soon as Director Li found out that he knew something about engines he hired him on the spot and sent him to Chahar to take charge of the plowing there. He had never seen a tractor but he learned fast.

—Chou Pu, the mechanic from the Pohai Gulf who had gone with me to Harbin. He was delighted to be chosen as a teacher, a position he had never dreamed of attaining. At every public meeting he got up to tell of his past, of the beatings and humiliations he had gone through to learn his trade. The only trouble was that he did not know when to stop and Director Li had to urge him to sit down in order to go on with the program.

—Lai Peng-nien, from East China. He was one of the original UNRRA students from the Yellow River flooded area. From there he had gone to Shanghai for a degree in agricultural engineering. He came to Double Bridge from a tractor farm in Kiangsu run by the East China government. He came to study Soviet machinery. Director Li made him a teacher.

—Four engineers who were interested in the manufacture of farm implements came from the Yangtze Valley. Someone in the Ministry got mixed up and sent them to the tractor school instead of the Farm Implement Factory and before they had realized the mistake Director Li had made them into teachers. It was not until the course was over that they finally got around to visiting the Farm Machinery Plant at Nine Dragon Mountain. They did not complain, however. Double Bridge was the only place south of the Great Wall where Soviet machinery could be studied and everyone knew that this was the machinery of the future.

I was made head of this teaching group. The teachers' main task was to keep ahead of the students. As soon as supper was over we rushed to the tractor shed to take the machines apart and study the problems that were to be taught the next day. Each "teacher" studied the parts allotted to him for an hour or so and then we met to pool ideas and discuss any questions that remained. Since we also had to answer all the questions raised by the students the day before we usually got to bed long after midnight.

Keeping at least twenty-four hours ahead of the students was difficult. So was teaching in the crowded classrooms. The only building large enough to hold both the tractors and the students was the former telephone exchange building. It was roomy enough, but there were no partitions. Each day four persons lectured and each lecture was given four times. The students, divided into four groups, rotated from one teacher to the next. As a mark of progress over South Ridge days, each student had a little stool to sit on instead of a brick or a block of wood. At the end of an hour they picked up their stools and moved to another corner of the room.

But regardless of which corner they settled into, it was hard to hear the teacher because there were always three other lectures going on at the same time. No sooner would one teacher pause in his explanations of the cooling system than a lucid description of the oil pump would come from another corner of the room.

We teachers in turn found our voices rising higher and higher in an effort to make ourselves heard. Ten minutes after the lectures started we found ourselves shouting at the top of our voices like barkers at a county fair. Then, realizing that this was impossible, we would stop the class and start in again pianissimo.

There were no doors on the building and no glass in the windows. The temperature inside was the same as that out of doors. Luckily the weather was kind that winter. A good warm sun shone down upon us day after day, and the force of the wind, when it blew, was broken by the high brick walls. Within a few weeks we had studied the whole tractor, from crank to drawbar.

While this was going on in the classrooms, Chiao Shih-ju led the mechanics in the shop in the repair of the tractor engines that had come in from the farms. By studying together they solved many problems that had seemed insurmountable. Their most brilliant achievement was to devise a substitute for the thin steel cylinder sleeves that had come with the Fords and had already worn out of round. These sleeves were so hard that they could not be bored out by ordinary cylinder boring equipment; even special cutting tools from Germany only scratched the surface. We had no spares. Chiao and his crew decided to replace them with cast iron. As they were almost paper-thin, all the experts said it would not work—cast iron could not be made than thin without cracking. But in desperation the mechanics went ahead anyway. They cast thick iron sleeves, forced them into the block, and then bored them out to size. The first engine that went on the testing stand ran very well. This was no indication of how it would stand up in the field, but since there was no alternative, they decided to gamble. All the tractors were fitted with cast iron. It worked. Only one tractor broke down that year because of sleeve trouble.

The winter was a success all around. Not only did the drivers go back to the farms confident that they could handle the Soviet tractors, but they also went back with a new spirit of solidarity. They had come from different regions with widely varied backgrounds. During the first few weeks they stayed in small cliques and were very conscious of their differences. But the course soon united them with a common experience and a common goal. The slogan in great white letters on the walls of the school, "We are the pioneers of the wastelands," took on a new meaning as the students realized that the same forces were at work in many different parts of the country simultaneously, and that the problems they faced they faced in common. Out of that training course a spirit of cooperation was born. "Tractor hands," north, south, east, and west, became brothers in fact as well as in name. One concrete result of this was the founding of a new magazine—the *Mechanized Agriculture Newsletter*—that published news from farms across the country and related the most up-to-date technical experiences to every field and shop.

Out of that course was born the friendly competition between farms that helped to improve the level of work everywhere. Each group left for home determined to make its farm the best in the country.

To the engineers from the Yangtze Valley, fresh from years of decadent Nationalist rule, the winter at Double Bridge was a revelation. One of them told me that he was returning south a new man. Under the old regime he had never taken any interest in politics. He stuck to his machines and his handbooks.

When Chiang was in power, he worked under Chiang. When the government changed, he worked on as before, convinced that one government was much like another and that ordinary people got a raw deal in any case. The thing to do was to find your own niche and mind your own business. But after spending the winter at Double Bridge he began to see things differently. "Are you glad you came?" I asked him.

"I certainly am," he answered.

"But since you came to learn production methods I should think this would have been a waste of your time."

"Not at all. We have learned about Soviet tractors. That's all to the good. But that isn't the main thing. The main thing is that we've had our eyes opened about China and about the future. Our talks with Director Li and all the discussions here have helped me to see that something new is really happening. I never thought Communists, or anyone else, could change things in China. I thought it was all talk. But things are already changing. Director Li and the people here mean what they say. They are going to build something new and I believe they will succeed. I hope we can carry that spirit back with us. It is a lot more important than all the technical knowledge in the world."

On graduation day the students and staff turned out in the brand new uniforms that were to become traditional for tractor workers in China. Blue overalls, strongly reinforced at the seat and knees, blue jackets that zipped up the front, and blue workers' caps. As they assembled in the tin-roofed shed that was the general meeting place, they looked like the ranks of a new and powerful army.

Director Li was glowing with pride. He wore the same outfit as the rest, and, as usual, it was a little too big for him. The pants bagged at the seat, the cuffs scuffed the ground, but he paid no attention as he hurried from place to place making sure that everything was in order.

All at once the whole group burst into song:

> We workers have strength
> Night and day we are busy
> Our eyes are red with weariness
> Our faces run with salty sweat

> Why, why do we work so hard?
> To liberate the whole of China,
> To liberate the whole of China!

This was followed by:

> Unity is strength.
> Iron is strong.
> Steel is strong.
> But unity is harder than iron.
> Unity is tougher than steel . . .

Their voices filled the air until it seemed that the pressure of sound would surely lift the roof off the shed.

Secretary Chang of the central Ministry of Agriculture spoke. "Our job is to open up wasteland, but where are the wastelands? They are not in the suburbs of Peking and Tientsin. We can't take a streetcar or a commuter's bus to work. No, the wastelands are far away on the edge of the Ninghsia Desert, across the Great Wall in Kalgan, buried in the swamps along the Yangtze, flung at the edge of the sea at the Yellow River's mouth. We cannot think of any easy life. We cannot hope for movies, opera, and soda pop. Someday we'll have everything. When we have created the conditions we can have whatever we desire. But this year and the next year will not be easy. Is there anybody here afraid of difficulties? Is anybody afraid of hardship?"

A great shout went up: "No! No! No!"

"That's right. We had better not fear hardship. It is a fact," he went on, "it is a fact that the imperialists are not afraid of anything. They go to the ends of the earth to seek wealth. Wherever there are resources lying in the ground or under it, sooner or later they come. Heat, cold, hunger, disease—they face it all for money. If we don't go out and reclaim our own land, if we don't go out and find our own oil and coal, and copper and tin, the imperialists will. In the modern world nothing can remain buried for long. No resource can remain untapped.

"Are we less brave than they? Could it be that we would do less for our country, our people, than they would do for gold? Could it be?"

Again the answer was a deafening shout: "No! No! No!"

"All right then," said Chang, laughing in spite of himself at their eagerness, "when you are assigned to your farms—some of them are far away, some do not even have village huts to live in—do not complain: 'Oh, I don't want to go there,' 'That place is too far away,' 'This place is too cold,' 'That place is too hot.' I hope everyone will gladly go wherever he or she is needed most."

"Never fear. We'll go wherever there is wasteland to plow."

They burst into song again.

It was the song that students all over China were singing that winter as they prepared to go out to the far ends of their land to start the great movement of reconstruction:

"Wherever the people need us, there we will go."

Chapter 21

A Trip to the South

MINISTER PAN, known to busy UNRRA officials as "Peter Pan," had been in charge of whatever "agriculture" the Nationalist Army still held when I had visited in Manchuria in 1947.

"The Communists," he used to say, "can't do anything for the peasants because they don't have any trained people. They have no one who understands agriculture."

"Don't you think technical people will work for a Communist minister once he takes office, just as they work for you now?" I asked him.

"What an absurd idea," snorted Pan.

Minister Pan fled to Formosa when Shanghai fell, but very few technicians followed him. I thought of this as I looked at the people in the Spring Sowing Inspection Team that Director Li was sending south to see how the men and women we had worked so hard to train were making out on the wasteland. We were going to visit our students-turned-drivers at South Ridge, Poai, and Yungnien. We hoped to help solve any problems which they might have run into, whether administrative or technical.

Our group included Lai Peng-nien, the "teacher" from Kiangsu; Li Mo-chiu, a technician from a private Peking garage; Liu, an economics student from the Management Division of the bureau; Peng, a raw-boned, loud-mouthed agricultural expert who had worked many years for the Japanese on experimental farms; and Chen, a Hunanese Agricultural College graduate, who was a member of the Democratic League.

They were all landlords' sons or the sons of men with land-lord connections. They came from the class which the Revolution had overthrown, yet they were not only working for the new government, but, with the possible exception of Li, were inspired by the great tasks of reconstruction. Already the People's Government had won from them a respect and loyalty that the Kuomintang had never engendered. In the past they had worked for salaries and whatever could be squeezed on the side. Now they worked for the progress of their country, and took its problems to heart.

We had not been on the train more than half an hour when Li suggested that we go to the diner and have some beer. We made our way back through the crowded cars and gulped down six bottles of "Five Star," Peking's best-known brew. To pass the time we then ordered cakes, sat around talking in loud voices, and finally washed down the last of the cakes with more beer.

When we got back to the coach Lai said: "I don't think we ought to live like that."

"Why?" I asked.

"It makes a bad impression."

"The trip will be hard enough later on. Why not play a little while we can?" asked Li.

"It isn't suitable for government workers to sit around in the dining car, drink beer, and spend money like a group of profligate merchants. In Peking we live simply. We are frugal and hardworking. We ought to do the same on the road. We ought to set a good example to everyone," Lai said.

"What does it matter as long as the money we spend is our own?" asked Li, who was treasurer for the group. He felt that if he kept public and private funds separate, that was enough.

"Lao Lai is right," Peng said. "Look at the peasant comrades on the train. They buy their food at the stalls along the way. We should do the same. If we want to be close to the people we cannot live in a different way and spend a lot of money on beer and wine. And besides, I don't know about the rest of you, but I can't afford it."

I agreed and so did Liu and Chen. We decided we had started off on the wrong foot and agreed to act like the conscientious inspection team that we hoped to be.

The question of study came up next. At the Ministry in Peking and at the training classes in Double Bridge everyone had studied

at least an hour every day before work began in the office or
the classroom. Were team members to fall behind just because
they were traveling? Once again Li Mo-chiu didn't seem to care,
but the rest wanted to keep up with their study and elected Lai
to lead them. Fortunately he had brought the books along.
Whenever we had a chance to sit down, read, and discuss we
took advantage of it. The topic for study that month was the
reorganization of the national financial system, the unification
of tax collection offices, the bringing together of all tax grain
in warehouses managed by the central authorities. It was a sweeping reform that would cut waste and undercut chances of graft
and corruption. If successfully carried out it could help stop
inflation, put China's finances on a sound footing, and lay the
basis for the nationwide construction work that was planned.

As workers in agriculture we were, of course, directly concerned with this progress. The financing of state farms and the
disposal of state farm crops were directly affected. The new
regulations put all state farms on a strictly commercial basis.
They could no longer live on their own resources. The produce
they raised had to be turned over *in toto* to the government.
The money the farms needed for operating expenses and for
investment was then allocated according to the plan and was
spent only if the plan called for it. Full grain bins could no
longer be considered cash in hand. The change meant that much
more careful management and intricate planning would be necessary. It also meant that the farms had to keep strict cost accounts
and conduct operations in a businesslike way. From that time
on every penny had to be accounted for: before a second penny
was allocated, the use of the first had to be justified.

Our first destination was Poai Farm, down near the Yellow
River. We had to travel to Hsinhsiang and then change to a
branch-line railroad that ran west to Chiaotso, a coal mining
center tucked under the edge of the Taihang mountains. The
branch-line train had no regular coaches and we rode in boxcars
equipped with benches. We passed through some of the most
beautiful scenery in the world, or so it seemed to me that
morning.

To the north the jagged peaks of the Taihang range blocked the sky. The villages that dotted the plain were ringed about with bright green squares of wheat, broken irregularly by fields of yellow flax in full bloom. Looking southward, the villages. fields, trees, and paths blurred together in a haze that we knew must harbor the Yellow River itself, for people said that the blue range of mountains on the far horizon was in Honan, across the stream.

As we traveled westward the land began to rise and undulate. The solid expanse of fields gave way to patches of orchard trees, some of them in full bloom. There were persimmons, cherries, pears, peaches, and walnuts. We could see that the upper branches of many of these trees had been grafted onto stock of a different kind. Fruit growing had been developed into a high art and its skills were passed on from one generation to the next.

The sun shone warmly down out of a clear sky. Prosperous peasants in bright new clothes climbed onto the train with bags of grain, chickens, and strings of dried fruit. They were on their way to the market at Chiaotso and talked animatedly among themselves. Spring had suffused the whole countryside with activity.

When we left the train at the end of the line we were still more than twenty miles from Poai Farm. We hired a cart to carry our bedding and set out on foot. Our road first skirted several large coal mines. Smoke poured from the chimney of a power plant. We could hear the rumble of the elevator machinery and the whine of steel wheels on a curved track as an engine shuttled empty gondola cars into the mine yard and hauled out a fresh trainload of coal.

Then the mountains crowded in from the north and orchards covered the land on every side. We crossed streams of sparkling mountain water that branched into irrigation ditches laid among the trees. These streams came from a mountain river down an irrigation system centuries old. A dyke tapped the torrent as it burst from the hills and sent the water outward like the ribs of a fan to service thousands of acres of land.

The sun was disappearing behind the mountains when we finally came to the farm. In the soft light of dusk the wheat fields, which had been planted in the fall, looked like flat sheets of emerald. Clear water gurgled and danced in the ditches on both sides of the road. Leaning willow trees sheltered the villages and settlements in the low-lying basin while on every side heavy thickets of bamboo, a rare growth in North China, cut off the wind. Only the highest mountain summits still peered over the tops of the vegetation to remind us that we had not suddenly dropped into a wholly new world. How rich and moist and fertile it seemed after the dry, windswept North China plain.

Mountains, rivers, running streams, a bamboo-sheltered basin of black soil—Poai looked to us like a garden of Eden set down in a forgotten valley.

We soon learned that the place was no paradise. The low wet land was a breeding ground for mosquitoes. Malaria tormented the villages. No one who settled here escaped for long the painful chills and fever. The tractor drivers told us that as many as a third of their number had been laid off work in one week due to illness.

The drinking water, too, was unwholesome. For some reason the people did not drink from the running streams but used only the water from shallow wells. The well water was full of minerals that caused diarrhea in newcomers and stiffened the joints of old timers. "With Poai water slake your thirst/Your legs by devils will be cursed," was a local saying.

One of the young mechanics from Peking, homesick and lean with dysentery, thought that it was not only his legs the devils were interested in. Devils came into his room at night, he said. He locked the door but they jumped in the window. He slammed the window shut but they crawled under the bed. He jumped onto the bed but they grabbed at his legs. He ended up screaming in terror, "Drive them away, drive them away."

When his companions broke into the room to see what was the matter they found him cowering under his quilt. "Help, drive the devils off, I want to go back to Peking, I want to go back to Peking," he shouted over and over again. For several days they could not leave him alone, so frightened was he of the stillness of the country nights, of the ghosts and devils described by the peasants, and of the sickness that he saw all around him.

When he returned to normal he still wanted to go back to Peking. It was clear to us then that pioneering had its casualties as well as its victories.

When we saw the land we realized why Poai drivers had welcomed the Soviet crawler tractors with such enthusiasm. The wastelands they had to plow were full of hummocks and hollows that stalled the wheel tractors before they could get started. The roots underground were enormous. Some were as large as a man's wrist and so matted together that the furrows did not turn over and lie down as they should, but broke up, stood on end, fell backward, and twisted sideways until the whole field looked as if a herd of wild boars had been let loose. Even the big Soviet NATI's couldn't pull the five plow bottoms they were rated for, or even four for that matter. The drivers had to cut the plows down to three bottoms in order to turn the land at all. Two or three years would be needed to convert that soil into mellow ground.

For some reason that we could not understand, machines went awry as readily as men at Poai. One after another the cylinder heads cracked. Was it the hard water? Or was it the strain of breaking heavy roots, wallowing over hummocks, and sinking in ditches? Or were the tractors poorly designed for the warm climate? The problem was serious, for there were very few spare heads in China. The cracks in the iron had to be drilled and patched, but the patching did not last. Plows also broke in that rough ground. When the moldboards snapped in two they had to be taken to the mines as Chiaotso for welding. All these troubles had held up the spring plowing and the farm was already a week behind.

As if disease and breakdowns were not enough, there were also personnel problems. Some of the drivers had been on the farm a year, others had just come from Honan, still others were new arrivals from North China. These different groups did not get on well in the field. They were jealous of each other. The newcomers thought they were just as competent as the old hands, but the latter had been appointed brigade and team leaders because the farm manager knew them better.

There was even more disunity among the mechanics. There

were no real experts—all of them had been chosen from the repair crews at Double Bridge. Not one of them would listen to what another had to say. Each felt he was as competent as, if not more competent than, the rest. So each went his separate way and secretly rejoiced when another failed to solve some problem, on the chance that this might enhance his own prestige.

Speeches about working together were no solution. What was needed was a man who knew enough about machines to win the respect of the rest and lead them to work together. The technician in charge did not have that much knowledge. He was one of the Harbin students who had been a truck driver, not a mechanic and his qualifications as a technician consisted of a year's course in the training school at Harbin and a few months' experience on a northeastern farm. As far as the mechanics were concerned he was an amateur. They thought he didn't know any more than they did, so they were not willing to listen to him.

To add to the difficulties created by friction among the workers themselves, there was a growing gulf between the management of the farm and the staff at all levels.

The farm manager was a man named Hsu. Before the Japanese war he had been a middle-school teacher. After the invasion of North China he joined the Communist-led guerrillas and become the general affairs secretary of his county government. In many ways he seemed an ideal man to head up a large enterprise. Although he knew very little about agriculture he had had years of administrative experience. He was level-headed, hardworking, honest to the tips of his toes, and extremely loyal to the ideals of the revolution.

What Hsu lacked was imagination and human warmth. Like a schoolmaster he weighed and considered every question and delayed action until the last minute. He rarely consulted with others and those under him felt frustrated. Decisions made without their participation they carried out only half-heartedly.

Two problems which troubled most farms were acute at Poai: food and medical care. On both scores we found many people grumbling. They said they were not getting the food they paid for. They ate coarse grain and cabbage day after day at prices that would have bought noodles and steamed bread in Peking.

As for the medical care—the sick had to pay for their own drugs and there was no doctor, only a man with limited first aid experience. Since everyone got sick at Poai, the workers felt that medicine at least should be free. They were upset because Manager Hsu's wife, a chronic invalid, got a subsidy from the provincial government for medical expenses. "The cadres are taken care of," they said, "so why shouldn't we be?"

Afraid of getting only the manager's side of the picture, the members of the Spring Sowing Inspection Group avoided Hsu entirely. We spent our time talking to the workers. When we heard that the food was not good we skipped the special breakfast that had been prepared for us and went out to the courtyard to eat with everyone else. We went to the shops and to the fields and everywhere asked what grievances people had. The more information we gathered, the more shocked we were.

After two days we had recorded a long list of problems. At a meeting attended by Hsu, the vice-manager, the secretary, and the brigade leaders, we read out the list as if it were a catalogue of charges against them.

Hsu looked dazed. He admitted many things were wrong, but he said that there were a lot of factors we had not considered. The discussion was inconclusive. We left the meeting feeling frustrated, and so did he. Hsu hardly seemed capable of managing so large a farm. It did not occur to us to have a long frank talk with him and find out how things looked from his point of view.

On the day that we left, as we walked eastward from the farm toward Poai town, I found myself alone with Hsu. I asked him the question that had been bothering me.

"What about this subsidy for your wife?"

"They told you about that?"

"Yes. There seems to be a lot of resentment over it."

"They don't know the background. Yes, my wife does get medical aid. It is because she ruined her health during the war. She was a leader in the guerrilla movement. She worked herself to exhaustion. The food was poor. They had to be on the move all the time. She finally had a breakdown. While the war lasted there was no rest for anyone. Her recovery was delayed. Now she is able to get some of the care she needs."

"Then it isn't because she is your wife?"

"No, of course not."

There was a long silence as this sank in.

"I don't know," Hsu said finally with a sigh. "I think this job is too much for me. I haven't been able to win the support of the men. I have had no experience with machines. I have decided to ask the ministry to relieve me."

All at once I realized what the effect of our visit had been on him. I tried to speak a few words of encouragement, but it was late for that. At the gate of the county seat he said goodbye. He had business with the district chairman. He walked down the street with his shoulders drooping and his head bent forward, like a man whose burdens are too heavy to bear.

On the train traveling back north, I asked the others if they thought we had done the right thing at Poai. Had we accomplished our mission? Had we left the farm more united, more vigorous, and more confident than when we arrived? To all this the answer seemed to be: No. We had been too one-sided. We had listened to the workers but not to the managers. We had catalogued the defects but not the achievements. We had understood part of the problem but not the problem as a whole. We had made no distinction between just and unjust complaints —as in the case of Hsu's wife—and we had laid all the complaints at Hsu's door as if he alone were responsible. The result was to lower, not raise morale, and to so discourage the manager himself that he asked to be relieved. Could the problems of mechanized farming be solved by encouraging all the managers to resign?

When Hsu said there were many factors we had not considered, he was right. The men wanted a good doctor, but could a good doctor easily be found? In the whole of China there were only a few thousand good doctors. The men said food cost more than in Peking. That was true, but it was not simply poor management. Although grain was cheap enough on the local market, all kinds of vegetables were expensive, including cabbage. There were no large market gardens around the county town of Poai.

The real problem was one of confidence and cooperation. They had to learn how to pull together. Then all the problems could eventually be solved. But we had taken the side of the workers against the management and helped to create an even greater rift. Had we not failed to grasp that Poai Farm was a socialist enterprise, an enterprise in which there should be no basic conflict of interest between the manager and the workers? In reality, in such an enterprise could there be any "side" of the workers? Could there be any "side" of management? Perhaps there could only be the "side" of how best to improve production so that the life of the whole people might be improved. Where exploitation no longer existed, was everyone not on the same side? If a manager had shortcomings, didn't the question become how could he be helped to overcome them? Were men capable of running ten-thousand-acre farms growing on trees? If not, then didn't the best men available have to be chosen, and then aided, encouraged, and cultivated until they became capable of handling the job? Was it any solution to mobilize their workers against them?

These were some of the questions we discussed as we traveled away from Poai. By the time we reached Yungnien, the next farm on our itinerary, we had decided to approach the whole question differently.

Yungnien Farm had changed a lot since October, when Chiao Shih-ju and I had gone there hunting gasoline. At that time a school for captured Nationalist officers had occupied most of the empty buildings. Now the officers had completed their studies and moved away to take up civilian work and the farm had expanded into their former quarters, which consisted of dozens of interconnected courtyards. The first time we walked there alone we got lost in the many different levels, alleys, archways, gates, and yards that separated the farm office and the repair shop. One wrong turn and we wandered into an unknown maze.

All these abandoned buildings were a grim reminder of the bloody fighting that had taken place during one of the longest and bitterest sieges of the civil war. A group of former bandit troops, affiliated with the Kuomintang after the Japanese surrender, had taken over the city of Yungnien in 1945 and held it against all comers. In 1947 the people's militia laid siege to the city and cut the enemy soldiers off from supplies. Chiang supported his garrison with air-dropped weapons, ammunition, and food. The soldiers lived a luxurious life—each one was reported to have several wives—but the civilian populace soon ran out of grain. People died by the hundreds in the streets and their bodies were thrown into the moat that ringed the town. So many bodies rotted there that the stench carried far into the countryside and the fish, feeding on human flesh, grew to unprecedented size. In the grand effort to take the city, the peasants of the surrounding countryside threw up a circular dyke and diverted the waters of the Wei River into the lowlands around the town. The water, rising slowly against the city gates, would have put an end to the garrison's resistance had not the Kuomintang commander radioed for air support. A plane flew in from Peking and blew up the dyke. The resulting flood lashed back at its creators, destroyed the crops for many miles around, and drove tens of thousands from their homes. After that a long stalemate ensued: the militia were not strong enough to take the town; the garrison was not strong enough to break out. Townspeople continued to die and the fish in the moat, which by that time had become a vast lake, grew larger and larger. Finally, when all the remaining Kuomintang forces in North China had been defeated, the Yungnien detachment made a desperate effort to break out and run for the South. The men did not get very far. One by one their small groups were rounded up and brought in by the armed people of the surrounding counties.

The lands of Yungnien Farm were marked out on the lake bottom itself and ringed the town in the shape of a doughnut. The extraordinary fertility of the rick black soil was due, the people said, to the corpses of both fish and men that had rotted there, and indeed the place did have a grim and ghostly aspect. Great dark medieval battlements rose from the sunken plain and towered over the half-empty town. The water of the moat, which reflected these battlements on its surface, was so black and turgid that one could not see into it more than a few inches. On every side there was only an empty expanse in which the tractors were all but lost. How many years would it take to revive this place? Not many, we decided, when we saw the farm staff in action. Although the land, the crops, and the expected yields were not nearly as good as at Poai, everyone seemed cheerful, cooperative, and full of plans for the future. The food was excellent. Since the farm buildings were situated right in the heart of the town and the town was a county seat, there was no great medical problem. The county clinic was right next door. Nor was the staff isolated and cut off from all cultural life. The workers and their wives joined with the students of the local high school to produce plays and to sing in the chorus. Many wives found employment in the town and added to the family income. Study circles and an evening school were well attended.

The man who had organized all this was Farm Manager Liu, a fat, energetic cadre from the Provincial Production Department who had formerly run a large cotton ginning plant at Weihsien, forty miles to the east. When he was put in charge at Yungnien Farm, he had not given up the gin. It was attached to the farm as a subsidiary enterprise and he managed both at once.

Unlike Hsu, with his serious schoolmaster's ways, Liu had a knack for getting along with young people. He understood the sense of adventure that had brought them into mechanized farming, gave his young technicians and brigade leaders their heads, and backed them up with all the support he could muster. He also paid great attention to comfort and living standards. He headed a committee that actively supervised the kitchen and saw to it that drivers in the field got hot meals and hot water for drinking.

Everyone told us that Manager Liu really cared for the well-being of the staff. They said he himself had shouldered a carrying pole and hauled boiled water to the fields when there was no one else available to do it. He also checked night and day to see that everyone had enough to eat, warm clothes to wear

when the wind blew, and hats for protection from the sun. All
this made him a most popular manager and helped to unite the
whole farm into a team that worked together with enthusiam.

Things were going so well at the farm that there was no need
to stay long. We asked for a truck to take us to South Ridge
and pushed off in a heavy rainstorm.

Chapter 22

Big Yang Sits
Under the Pomegranate Tree

WE ARRIVED IN SOUTH RIDGE just as the wheat was making its
first spurt of growth. Kuo Hu-hsien's fears for the crop appeared
to be justified. Where the alkali was heavy the new sprouts
were already stunted and yellow at the base. Even the best wheat
was not as green and strong as that grown by the more skillful
of the peasant farmers. Nevertheless, everyone was hopeful. With
a few good rains, they said, a crop of fifteen to eighteen bushels
to the acre could certainly be harvested. Such a yield would
not be below the local average.

In spite of this optimism the morale of the tractor drivers was
low. Preparations were under way for heavy spring work, for
the planting of sorghum and cotton, for the plowing up of new
areas of wasteland—but the young people went about their tasks
with less than their usual vigor and spirit. I sensed that there
must be some friction between the tractor brigade and the farm
management, but it was not easy to get to the root of it.

After talking with a number of South Ridge people one thing
became clear. No one had confidence in Big Yang, the man who
had again assumed responsibility for the farm after Chang
Hsing-san left to set up the State Farm Management Bureau in
Peking.

Yang was an old man and very large. Everyone called him
Yang Pang-tzu (Big Yang). He seemed often to be in the grip
of a strange lethargy. He sat for long periods in one place,
usually a sunny spot, without moving so much as an eyelash.
At such times he apparently neither slept nor thought, but
simply existed, inhaled and exhaled, vegetated like a turnip.

Beside the door of the farm office grew a beautifully proportioned pomegranate tree. Sitting peacefully under this tree, his eyes half closed, Yang Pang-tzu looked for all the world like one of those carved mahogany buddhas that are so common in the curio markets of China. If he had donned a flowing robe in place of his padded cotton jacket and had crossed his legs under him the image would have been perfect.

Yang came from a solid middle peasant family with close ties to the soil. He had great prestige in his home region, which was not far to the south of Chihsien. During the anti-Japanese war he had joined the resistance movement and had been so well thought of by local leaders that he was sent all the way to Yenan to study resistance policies and guerrilla warfare. He walked several hundred miles through the mountains to get there.

Perhaps age and weariness had since overtaken him. Perhaps he felt he had done enough and it was time to relax. Perhaps he felt incapable of handling the complex problems posed by a large mechanized farm. Just what the trouble was it was hard to say, especially since Yang didn't talk much about it.

He was used to acre and half-acre lots the whole of which could be seen from the shade of one pomegranate tree. Since he did not have a car and could not ride a bicycle, he did not even get around regularly to look at the 2,000 acres under his care. Some of the best wheat grew not far from South Ridge village. It could be seen from the office gate. Yang rarely saw any other. His solution to alkalinity, as to almost every other soil and crop problem, was to plow, plow, and plow some more. He did not realize that a tractor plow that turned the soil over from six to eight inches deep was very different from the peasants' iron-tipped stick that scratched the surface like a cultivator. He had a pathetic faith that somehow, if the soil were turned and stirred often enough, the salt would fade away and fertility would rise from below.

Old Yang spent more time worrying about the garden outside the farm's back gate than he did about the whole rest of the farm put together. The garden was on a scale he could grasp. He personally supervised the construction of a wall which enclosed it

on all sides, and the digging of a well with which to water the
parched soil assured vegetables for the whole farm staff. Yang
relaxed and let nature have its way with the wheat.

It should not be supposed that Yang was a bad man. He was
not. He was a good man, a very kind man. He had served the
Chinese people and the Chinese Revolution well for many years.
Faced with a responsibility that demanded too much he clung to
what he knew while the rest of life rushed past. He lacked what
Chairman Mao called a "sense of the new," and so, inevitably, he
became a bureaucrat. He was not an autocratic bureaucrat who
ordered people around, or a *hsin hsin ku ku te* (busybody)
bureaucrat who wore himself and everyone else out looking after
minor details. He was a do-nothing bureaucrat who preferred
contemplation to action and continued to hope that the whole
world would slow down a little.

As the *People's Daily* pointed out time and again, bureaucracy
was the "hot bed" of waste and corruption. Irregularities flour-
ished around the honest, well-meaning, and often hardworking
bureaucrat in every field of work. Energetic and ambitious op-
portunists took advantage of the situation to pursue their own
ends. South Ridge was no exception.

Since Yang did not grasp the reins of control very tightly,
another man took over the day-by-day operation of the farm.
This man was Secretary Li. (Not to be confused with Li Chih,
director of the training class.) He was administrative assistant to
Yang. He was supposed to run the farm office, receive and send
letters, oversee the use of funds, and help the managers in every
way possible. Secretary Li did far more than that. He made most
of the decisions, disbursed funds, hired new staff members, called
and supervised meetings. His position was strengthened by the
fact that he was concurrently the chairman of the local Communist Party branch. Not only farm affairs, but Communist Party
affairs as well, were in his hands.

Secretary Li was a native of Nankung, the thriving cotton and
commercial center thirty miles to the south. For many years he
had been a cadre in the revolutionary government there, but his
thinking was still very close to that of a small tradesman. This
was not strange, for Nankung was famous for its merchants and

its trade. Every peasant family had at least one relative in business. In Nankung people were far more concerned with profit than with production.

When Secretary Li came to Chiheng he found huge sums of money waiting to be invested in crop production. These monies were the farm's operating capital. To see such funds sit idle, even temporarily, disturbed Secretary Li's entrepreneurial soul. Why let fertilizer funds gather dust until spring when they could be invested in trade and make a handsome profit for the farm in the meantime? Li chose the consumer co-op as the most likely place to put these funds to work. The co-op had been established by Director Li, head of the first tractor training class, to provide daily necessities for the staff and students of the center. Towels, toothbrushes, cigarettes, coal, notebooks, paper, pens, and ink were sold. Secretary Li immediately set to work to expand its operations. He turned it into a wholesale house. He bought cigarettes by the mule cartload, sold them to local peddlers *pi fa*—in bulk—and allowed them to make a sizeable profit marketing them retail. The co-op also became a coal distribution center for villages in a wide arc west of the basin. This trade also proved profitable.

To handle the expanded activities of the co-op, Li hired friends and relatives from Nankung. Soon the co-op staff outnumbered the administrative staff of the farm. With his own men in the majority, Secretary Li's wishes usually carried the day. A great part of every staff member's time and energy was spent discussing commercial operations.

Secretary Li was constantly looking for areas in which to expand further. Several small pump engines sitting in the stockroom gave him a bright idea. He set up a cotton ginning plant in Chihsien and contracted to gin government cotton at so much per bale. Mechanics were dispatched to keep the gin running. The sub-soil of the basin was good for making bricks. This gave Secretary Li another bright idea. He decided to go into the brick business. He located a manager in Nankung to set up a kiln. Twenty local peasants were hired in the off season as brickmakers. Truck transport was also lucrative. Secretary Li dispatched the farm trucks on commercial hauling expeditions.

In the meanwhile farm work fell into a pitiful state. I have
already described the wheat. The sorghum field was choked with
rank weeds. Scores of tons of sheep manure, which were to insure
a good cotton crop, still lay in the sheep pens where they had
accumulated over the years. The energy, the capital, the talents,
and the organizing ability allocated by the state to reclaim
marginal land and grow bumper crops were being diverted by
Secretary Li into subsidiary, non-agricultural enterprises which,
though profitable, had nothing to do with farming.

In this respect Secretary Li was only following in the footsteps
of Big Yang. He was sticking to what he knew best: in his case
this was not gardening but business. Profit and loss, margins,
rebates, retail and wholesale—Li had sucked all this in with his
mother's milk. Given capital to play with he was perfectly
capable of running it into a million. Secretary Li was bolstered
in his natural affinity for business by the formidable task of con-
verting the basin into a flourishing farm. He had very little faith
that any crop ventures could succeed in making money. Hence
he took the path of least resistance and expected that by showing
financial gains he would win favor with the bureau that employed
him.

In this he openly defied the People's Government. There were
strict regulations laid down for just such irregularities. Funds
allocated for production were to be used for that and that alone.
Trading, transport, brick kilns, cotton ginning—these had no place
on a state farm unless designed to supply its own members, process
its own products, or transport its own supplies. Business for business' sake was illegal. *Chuan kuan, chuan yung* (earmarked funds
for earmarked use) was the slogan. But the country was large,
enterprises were many, a clever manager could distort national
policy for a fairly lengthy period without being corrected.

It was the distortion of government policy that upset the
tractor drivers. They had come to build a farm, only to find themselves submerged in multiple commercial enterprises. With Secretary Li business came first, tractors second or third. He fired
Liu, the blacksmith, diverted skilled men to the cotton gin when
the tractors needed their attention, and used the tractors to haul

bricks. Everyone was angry, but there was not much they could do about it.

The diversion of funds turned out to be only a part of the story. One night Lao Kuo, the former militiaman from the Taihang, asked me to come to his room. "I want to tell you something about finances," he said.

When I arrived he told me that the drivers suspected Secretary Li of corruption.

"Did you see that cotton gin in town?" he asked.

"Yes."

"Well, every month the State Cotton Company gives the gin so much cotton to process. The cotton is top grade. We are supposed to return top grade cotton. But Secretary Li, in league with a Nankung merchant, sells a portion of the top grade cotton, substitutes second or third rate stuff in its place, and turns the mixture back to the government. He himself pockets the difference."

I was stunned. "Are you sure about this? Have you any proof?"

"We can't prove it absolutely. But that is what those who work at the gin say, and there are other things which tie in with it. You know that Secretary Li gets a moderate salary just as the rest of us do. Yet he supports a very large family. There are at least eight people living off him here at the farm. They often eat meat dumplings, eggs, and pork. These are foods the rest of us can't afford. Where does Secretary Li get the money? He is the only one in the family who works. His wife has no job as many of ours do. Everything they have must come from him."

"Have you reported this?" I asked. "Couldn't it be investigated?"

"We took it up with the Provincial Agricultural Department but they have done nothing. They said they would send someone down to investigate but no one has come. We also took it up with the Party branch. But Li is chairman, he denies everything."

"Can't you appeal to a higher body?"

"I have. I went to the regional committee. But they have known Li for a long time and can't believe he would do such a thing. Also, he is a local man. Chi Feng-ying (one of the three girl students of the first training class who had stayed on at Chiheng as a brigade leader) and I are outsiders. They think we may be trying to discredit him."

"But surely they will send someone to investigate."

"Maybe they will," said Kuo, "but they haven't yet. They still wonder if I am not making this up for personal motives."

"What about the general farm meeting? Hasn't this matter come up there? Surely everyone knows of it by now?" I asked.

"We are outvoted on the farm council. In the general meeting many are reluctant to speak. Lao Kang has spoken out. He had a personal quarrel with Li. But he gets so angry that you can hardly tell what he is saying."

"What is his personal quarrel?'"

"It concerns his wife. You know, he just got out of the army. He married a widow, but she is not such a proper sort of person. Lao Kang is away a lot. He does the buying. He is gone for a week or two at a time. His wife has taken up with Li, or so he says. When he is away they carry on. Li has given her expensive presents, so now Lao Kang wants a divorce."

"But he only got married a few months ago."

"Yes, but he is fed up and angry. He doesn't want anything more to do with her. She really is not a good woman. But the marriage could be a success if there were no temptation."

"Ai ya!" I said. "This is fantastic."

The whole story made me feel sick through and through. It was bad enough to suspect any government cadre of corruption or immorality, but to have such a man at Chiheng Farm undermining morale, using government funds for personal gain, and allowing all the hard work of a year to slide into ruin in the meantime—this was a blow that struck home. Each of us who had worked at Chiheng felt a personal stake in the welfare of the farm. It was our first child—a hard luck child no doubt, but still a first child—and we were determined to give it every chance. Then along came a man who turned it into a personal gold mine. It was appalling.

I wondered if it were possible for one man to ruin what a hundred had built up. Could the drivers, young people like Kuo and Chi Feng-ying, ever allow it? All their training cried out against the corruption and perversion of their work. Wasn't the vision of a new China too bright to be tarnished and brought low by one selfish manipulator who tried to "drag them into the mud hole"? And yet their protest seemed to be gagged. The higher authorities were hesitating. Bureaucrats were passing responsibility from hand to hand. How could the problem be solved?

The group of technicians from Peking was not an investigating team. We had no authority to conduct an inquiry into graft and corruption. The best we could do was to raise the problem when we got back. We did, however, have the right to challenge the general policy of the farm. At a meeting of the farm council we sharply attacked the diversion of funds to projects which had nothing to do with agriculture and warned that the farm would be judged by the crops it harvested, not by the number of cases of cigarettes sold. It was obvious that if the funds were properly used graft would be far more difficult. There were no easy profits to be wrung from the wastelands. A good crop there could only be the result of hard work and careful planning, and that was not the kind of row that grafters liked to hoe.

We returned to Peking with heavy hearts, unaware that in the future a great movement against graft, corruption, and bureaucracy would expose men like Li all over the nation. In the meantime our job was to plan for the harvest, the first mechanized harvest in North China. For soon the wheat would be ripe and Li or no Li, we would have to harvest it.

Chapter 23

When the Sun Stands
One Stalk High

Bold young men and maidens fair
Fear not work, nama yi ya hei
Before the cock's crow fades away
We mount the combines to start a new day.

The tractor thunders o'er the field
The combines roars, nama yi ya hei
Watch the tractor roll and turn
Hear the combine whirl and churn.

Now the sun stands one stalk high
One stalk high, nama yi ya hei
We've harvested three tanks of grain
Grease up, fuel up, start again.

THE YOUNG MEN AND WOMEN in the railroad car sang lustily as
Hsueh Feng, now a dean of students at Double Bridge, made up
the words to verse after verse. On the line from Shihchiachuang to
Tehsien, boxcars with benches still served for passenger coaches,
so we were all crowded in together instead of being scattered
through the train wherever there was an empty seat. The group
was made up of technicians and drivers from farms all over North
China. There were a few from Honan, south of the Yellow River,
and one from Shantung. On that sunny afternoon as we rolled
along at a leisurely pace through the fields of ripening wheat and
plots of young millet and cotton the spirit of adventure soared
high. The other passengers smiled and exchanged curious glances
as they watched the young people in worker's garb sing out at the
top of their voices.

"Where are the comrades going?" asked an old peasant who held a pipe about two feet long and enjoyed two puffs of smoke for each laborious filling of it.

"Down to South Ridge to harvest the wheat," came the reply in chorus.

"And what would workers be doing at the wheat harvest?" asked a merchant in a long blue gown. His newly purchased bolts of cloth were stacked near the open door.

"We're going to harvest it with machines," said Lao Hei, the mechanic who had once offered to be my blood brother, with undisguised pride.

"Machines? What sort of machines are those?" asked the peasant, incredulous.

"Combine harvesters, from the Soviet elder brothers. They have knives sixteen feet long and can cut two hundred *mou* in a day."

"Is that a Soviet brother?" asked the old man, pointing me out with the stem of his pipe.

"Oh no," said several of the drivers at once. "That's Lao Han, our American technician."

"Can he run the machines?"

"Sure, of course he can," said the drivers, almost indignant at the question.

I wished that I felt so confident. I had never run a combine in my life. I had spent one afternoon years ago filling bags on a combine in an upstate New York oatfield, but aside from two dollars I had gained nothing from the experience. I hadn't even looked inside the threshing chamber. Yet here I was responsible for the operation of seven enormous Soviet harvesters. I kept wondering if the preparations had been complete. Did we have enough parts? Would the machinist from Peking be able to handle the mobile lathe shop that had just arrived? Was there enough oil? What if the canvasses got wet?

There was nothing that could be done about those questions now, but still I worried. Everything that had been done in South Ridge since the tractor class first opened—the training courses, the gasoline, oil, and seed that had been consumed, the patient work with the peasants, the renovation of buildings, the planning and the endless work—all would be on trial in the next two weeks.

The drivers did not share my anxiety. After a two weeks intensive course on combine operation in Double Bridge they felt they knew what the big machines were all about. After all, they had assembled them, learned the function of every working part, started them up, and driven them around. What had at first looked like a confused mass of steel, iron, chains, sprockets, reels, and levers had gradually taken on rational lines. There was the sickle knife that cut the standing grain, the revolving canvas that carried it into the box, the threshing cylinder that smashed the kernels from the heads, the straw walker that walked the stalks to the rear, the sieves and fans that separated the wheat from the chaff, blew the lighter material away, and let the heavy kernels themselves fall to the bottom. And finally, there were the elevators that carried clean grain to the top of the machine where it accumulated in an enormous tank until a truck could come to carry it away. Should anything go wrong with all this, the drivers thought, Lao Han would help them, as in the past, and everything would be solved.

I was not so sure. The fans could be made to turn at different speeds, the sieves to leap at different intensities, the apertures for wind and chaff could be widened or narrowed, the teeth of the cylinder set close or far apart. The combinations and permutations were endless. Only experience could dictate what was the proper way or the best way. Experience none of us had had.

The textbooks and instruction manuals were useless. The American books simply described the machines, told how much they could harvest in a day, and analyzed the cost per bushel. About adjustment, operation, wet conditions, dry conditions, wheat as distinct from rye or barley, they said not a word. The Russian books were in Russian, which none of us could read, and even if we had been able to, we would not have benefited much, for according to an interpreter, they dealt mainly with specifications and repair. One thing at least we had, a complete maintenance chart that showed how often each part should be greased or oiled or attended to. Since there were dozens of grease fittings we had invented a system of colored dots, red for four hours, green for ten, yellow for twenty, and so on. Each fitting had its dot and on the back of the machine the key was painted in large letters. Why hadn't the builders thought of that in the first place? I wondered.

"Lao Han, come on and sing," said Chang Ming, the tractor student with a middle-school education who aspired to be an engineer. He said it mockingly—he always liked to tease me by talking Chinese with a terrible accent which he said was a copy of mine. "What are you worrying about?"

"I was thinking about the combines," I said.

"Well, stop worrying. Before the machines came you were worried that we wouldn't have them and how would we harvest the wheat. And now that they are here you worry that we cannot use them. Never mind. Everything will be fine. Let's sing."

Chang Ming was right. I worried too much. All winter I had fretted over the Chiheng harvest. By dint of prodigious persuasive effort I had managed, while in Manchuria, to convince Bureau Head Ku to give us a threshing machine of enormous capacity, but what was one threshing machine for two thousand acres? And how was the crop to be cut and hauled? I did not believe that combines would really come from the Soviet Union in time for the summer harvest. I thought, of course, that in the long run machinery would come from Stalingrad and Kabarovsk, and the Altai, but that was something for the future. It had nothing to do with 1950. The harvest at South Ridge would have to be cut by hand. Mechanized farmers in the fields with sickles! Talk about "face." That would be a real loss of face.

Then suddenly, a month before harvesting time, the machinery had begun to pour in—trainloads of tractors, combines, grain drills, plows, and harrows arrived from the Soviet Union. None of us would soon forget the first night of that deluge.

We could not see the end of the long line of flatcars that stretched westward from Double Bridge station into the darkness. On each car sat two enormous tractors waiting for the crank. They had to be started up and driven off before dawn. In two hours we managed to move only a dozen of them and we were already exhausted.

The huge kerosene-burning monsters were stiff and sluggish. It was all we could do to crank them over, but cranking over slowly was not enough to start them up. Every now and then one would spring to life but most of the time we worked in vain. We put gasoline in the petcocks. The motor sputtered and died. We put gasoline in again only to have the same thing happen. There didn't seem to be any rhyme or reason to it. Some of the tractors just would not respond.

We tried cranking with two men. We tried cranking with ropes so that four or five could pull at once. Nothing worked.

About two in the morning Director Li disappeared in the jeep and came back with hot millet porridge and steamed bread. We squatted around the steaming pot under the glare of the electric lamp that lit the station platform. The task seemed hopeless. The great machines sat cold and silent on their cars. The station master came out to ask how we were progressing.

"There are lots more up there yet," he said.

"Yes, so it seems," said Director Li.

"Think you'll get them off in time?'"

"Doesn't look like it now. How about giving us more time?"

"Well, it's supposed to be four hours, after that you pay demurrage. But the engine won't be in until nine o'clock. If you have them off by that time we won't ask anything."

We started in again. The rest and the food had renewed our strength and for a while motors popped to life all up and down the line. But the vigor soon faded. We settled down to a long slow fight. Draining the fuel helped in some cases, but in others nothing seemed to work. We had to keep cranking even though our arms and legs ached beyond endurance and refused to obey our brains.

The sky grew light in the east. The sun rose. We could feel the growing heat on our backs but still the loaded cars kept coming. Finally, just as we heard the whistle of the engine on its way, the last tractor sputtered and caught and was driven off.

We started to climb into the truck that would take us back to the school when somebody shouted:

"Look, combines!!"

And sure enough, as we turned to watch, the engine glided into the station with a trainload of new machinery—great, gray, sheet-metal monsters sitting on flat cars with their wheels blocked and wired. C-6, said the faded white paint on the grain tanks. The

"C" stood for Stalinetz and the "6" for the six-meter-long sickle bar that filled the rest of each car.

Chiheng's harvest was safe at last!

We were so excited that we forgot about the agony of the night and set to work unloading the combines without stopping for breakfast.

Telegrams immediately went out to all the farms to send in their best drivers for a training course.

Hastily, technician Wan, who had once studied in America with the International Harvester Co., and who came to us from the University of Nanking, and I organized the study of the machines. It could hardly be called a course since there was no one qualified to teach it. As always we worked together, exchanged ideas and argued points of disagreement. Now we were on our way to the wheat fields, and new worries had replaced the old. That was life.

Hsueh Feng, who loved music far more than machines, or politics, or anything except possibly food, had already dreamed up a last verse to his song. As he repeated it over and over so that everyone could memorize it, I joined in. The verse was perhaps over optimistic. But why bother about that when everyone else was enjoying himself? I sang along at the top of my voice:

> The combines make a hula hula sound
> A hula hula sound, nama-yi ya hei
> The last grain of wheat has been hauled away
> Happily we sing at close of day.
> Hei, hei, hei, hei
> The last grain of wheat has been hauled away
> Happily we sing at the close of day.

Chapter 24

Combines in the Cotton Patch

AT THE STATION we were met by the farm carts and rode down through the familiar countryside under a clear sky that promised good harvest weather. The further south we went the poorer the wheat became. It had obviously been very dry. We became more and more anxious about the South Ridge crop and could hardly wait to see it. As soon as we came to the edge of Chiheng Farm, Hsueh Feng and I dropped off the cart and began a tour of the fields that lasted until nightfall. We found some fine stands of wheat with stalks thrusting up thigh high, but the more we looked the more certain we became that this was no bumper crop. As we had foreseen, the alkali spots were the worst; in many places there was no wheat at all, only the bare ground. This was especially true where the ground was high because every ridge and raised hummock increased evaporation and concentrated the salt. The wheat shoots died before they grew an inch tall. In other spots the wheat had come up well but had failed to grow vigorously and was now beginning to head out only a foot or so above the ground. The absurdity of the widely spaced rows in which it had been planted was clearly revealed. The growing wheat did not begin to cover the land. There was room between each thin green line for at least two more sets of plants. But there was nothing that could be done about that now; experiments with closer planting would have to wait for the next crop.

After surveying every field, we estimated that the crop would not even run ten bushels to the acre. There were twelve, fifteen, and even twenty bushel patches, but the average of the whole would not be over ten, if that. The only consolation was that some of the peasants in the surrounding area had even poorer stands; on the other hand, a lot of them had much better. With all our machinery, gasoline, deep plowing, and technical proficiency we were still behind the peasants.

We came across an old man picking bugs from some scraggly cotton plants.

"What do you think of the Chiheng crop?" Hsueh asked him.

"Very good," he said. "Very good."

"It's certainly not as good as yours," said Hsueh, wondering if the man was holding back the truth for fear of offending us. "How can you say it is very good?"

"That's wasteland out there. The salt ruined it ever since the flood. It has grown nothing. I think you got a good crop."

We found the same reaction everywhere. The peasants hadn't expected to see as much wheat as was growing there. This made us feel much better. Anyway, it was too late to do anything about it. Ten bushels or five, it still had to be cut and stored. For the next twenty days all attention was turned to the harvest. We thought, talked, and dreamed of nothing else. Big Yang's small peasant ways, Secretary Li's commercial deviations, his misuse of farm funds, the poor results in the cotton field, the fiasco of the sorghum, the overstaffing of the co-op—all these would have to wait. If the wheat was not cut on time it would be lost. And yet these problems turned up to plague us whether we thought of them or not.

We ran into trouble right away trying to find a place to assemble the combines. The wall that Manager Yang had built around the garden was supposed to encompass a tractor park as well, but Yang was so delighted with this enclosure that he had planted the whole area to cotton. Since the land was in first-class condition—it had been bought from the village only a few months before—the tractor-yard cotton was doing much better than the main crop in the basin.

It was the pride of Yang's life. He protested against moving the machines in and was deeply hurt when we insisted that there was no place else. When we finally had our way he suffered over each strong green plant that was trampled in the dirt.

Neither Yang nor Secretary Li had any faith in the ability of the combines. In America and in the Soviet Union combines might work. Certainly nobody would go to the trouble of making such enormous monsters if they couldn't do any work. But after all, this was China. Maybe Chinese wheat was different. It had always been cut by hand. It shattered easily. Also, the crop was poor. Maybe it was too low. There wasn't much straw. Wouldn't the wheat all be blown out behind? And besides, could the young drivers use the machines? Would they be able to adjust them? Yang had read about a combine fiasco in the Northeast, where a farm had tried to use combines for harvesting and had lost a major proportion of the crop. If there were delays it would be too late to hire local people. Then everything would be lost.

Secretary Li thought it best to take no chances. He got in touch with the local high school and the Teacher Training Institute and asked if they would send students to help. He also arranged with the commander of the district garrison to send men down to guard the crop. If worse came to worse they could help out.

Fortunately for Chiheng Farm, Manager Liu came up from Yungnien with his tractor crews and under a directive from the Provincial Agricultural Bureau took over command of the harvest.

He proved to be an energetic organizer, full of enthusiasm for machinery and new ways. He did not stop to worry about everything that might go wrong. Instead he went right ahead and plunged in. "You can't learn to swim if you don't jump into the water," he said.

Liu organized the harvest campaign as if he were going into battle, which was exactly the way we all felt. The combine crews were the front line fighters. They had to be supported by supply contingents, reinforced with mechanics, and protected by first aid men. Fuel and parts had to be shifted to the field, hot meals sent out, and boiled water supplied at all times. The harvested grain had to be hauled home, dried on prepared ground, cleaned, weighed, and stored. Something had to be done about the peasants, some of whom had already taken advantage of moonless nights to carry off a *mou* or two of the best wheat.

Since the farm was divided into two main areas which were more than four miles apart, all these activities had to be set up in duplicate. Two field kitchens were necessary, and two mechanic's teams, and two medical kits. There was no way to divide the

mobile machine shop, so it was assigned to the farthest section
and set up operations at the sheep pens east of Han Family Village.
Liu and Yang divided the harvest area between them and took
turns presiding over first one side and then the other.

I was the only one who was expected to be everywhere at once.
Since I couldn't be divided in two, I was given a new Shanghai-
made bicycle with which to pedal furiously back and forth. An
outstanding feature of this bicycle was a beautiful electric generator and light imported from Holland. The peasants in the villages
along my route were delighted with it. They had never seen the
like and when word got around that Teacher Han's bicycle *mo
zu chi te tien* (ground its own electricity) I had to stop at frequent intervals along the way to demonstrate it.

A week of tinkering, tightening, repairing, and adjusting
finally put the combines into shape. The weather held fair and hot,
the wheat ripened quickly. The peasants were already busy hauling in the sheaves and stacking them around the village threshing
floors. It was time to begin.

As the great machines went lumbering out of the gate with
the Soviet tractors in front blasting their exhaust to the sky and
the headers trailing behind on twenty-foot wheeled racks, it
looked as if some long disjointed Chinese dragons had suddenly
come to life. At the last minute the drivers had somewhere unearthed red banners, and from the highest point of each combine,
front and rear, they streamed out proudly in the wind. A great
cheer went up from those who stood watching. The combine
crews cheered back. Only Big Yang looked unhappy; the best
of his cotton plants had been destroyed. As the noise of the
machines faded in the distance he looked the garden over carefully and shook his head. The cotton would never recover.

Chapter 25

It Works

BANNERS AND CHEERS were all very well but now we were face
to face with the ripe wheat itself. It stretched out on every side,
bending with the breeze and rustling quietly—peaceful, serene,
provocative. Would the combines succeed? Could they really
digest this enormous expanse? I felt as if I had had nothing to
eat for several days.

For a beginning we selected a good stand of wheat east of Han
Family Village. I told three of the crews to wait while I went
with the fourth to make the first cut. Lao Hei started the engine,
let the clutch in slowly, and brought the whole rattling mechanism
up to speed. It shook, whined, and roared, but nothing gave way.
So far so good. We had done that much before. Then Lao Hei
pulled a long blast on the exhaust whistle; the tractor started up
with a jolt and the great long header plunged into the wheat.

I don't know what we expected to happen. Perhaps the knives
would plug, or the wheat would fall on the ground, or the canvas
would jam—but no such thing occurred. Urged by the sweeping
reel, the golden ears came dancing onto the canvas. The canvas in
turn bore them rapidly to the cylinder that knocked them to
pieces with a low snarl. Soon straw began to bounce from the
rear of the combine, and then grain began to pour into the grain
tank. We watched the growing stream full of wonder and excitement. We took turns leaning over to hold our hands under it. The
wheat poured over our palms, bounced from our wrists, and
fell with a drumming sound to the metal below.

"It works!" said Lao Hei, jubilant.

"Look how clean it is," shouted Chi Feng-ying. "As if it had
been winnowed ten times."

I noticed some chaff and bits of straw slithering down the chute along with the grain, but this looked like a minor problem. The machine was certainly working.

Already we were rounding a corner. For a moment the flow ceased, then started up again with a rush as we bore down a long straight stretch. We listened for any false notes in the motor. There were none. We climbed down the ladder, squatted on the main header brace, and watched the wooden rod that drove the knife. It flashed back and forth in a blur, but it made no alarming noise. All was well there. We opened the inspection hatch and watched the straw coming up the conveyor; we inspected the sieves shaking the chaff to the rear. We jumped to the ground and ran to the far side of the machine to examine chains and sprockets. Everything was in order. In the din created by the spinning, shaking, bouncing parts it was almost impossible to talk. But at that moment no words were needed. We looked at each other and smiled.

By the time our combine rounded the end of the field and started back, the other machines were no longer in sight. Impatient with waiting, they had started harvesting on their own. When the machine I was on completed a full circle I jumped off, waved the crew on, and stood a minute looking around. Like great gray battleships, the three tractor-drawn combines moved through the wheat while over on the left the self-propelled machine rolled across the horizon like a light bomber about to take off. The roar of the engines sweeping across the wheat from all sides spoke triumphantly of the tremendous power that had suddenly been unleashed.

I turned to Manager Liu, who stood there transfixed.

"If they go on at that rate we'll be finished up in two or three days."

I had no sooner said these words than one of the combines came to a halt. Then all four were standing still in the field. The throb of the motors ceased entirely. The mishaps had begun. One of the combines had a broken elevator chain, another had lost its drive belt, the canvas had jammed on the third. After an hour of hard work, running from one to the other, I finally got them moving only to find the same process starting over again.

During one brief moment when they were all running I peddled off to South Ridge to see how the other brigade was making out. Contrary to instructions they had not waited for me. They had long since set their machines running, blown a blast on their whistles, and plunged in. Now they too were stalled all over the field with minor breakdowns.

At first I was angry, but then I thought it was just as well. If we waited to get all the kinks out of one machine before starting the rest the grain would be falling to the ground. It was best for all crews to jump in and begin. That was the only way to learn.

And learn we did, the hard way.

One crew left a rope on top of their machine. It fell off, caught in a fast moving chain, jammed a sprocket, and smashed it in two. There were no spares. Another crew ran their header into a parked tractor and crumpled it. A third forgot to tighten the bolts that held the main screens. They shook themselves loose and wore the fastenings so that it was impossible to tighten them later. Canvasses jammed, ripped off their slats, and snapped their straps. Sickle knives came loose and broke guards. Even the tractors refused to behave. One blew a gasket. Another shook a head bolt loose. It seemed to me that everything that could go wrong went wrong. The expanse of uncut wheat loomed larger and larger. In three full days we had harvested less than a fifth of the area. The peasant's crops had long since been out and hauled home. Soon the wheat would begin to shatter. Time pressed like a suffocating blanket building tension far beyond anything we had experienced in the spring.

To make matters worse I found that the wheat dust affected my head, throat, and lungs. As soon as I got near a combine I began to cough and sneeze. After a day in the fields I could not sleep at night. My throat seemed to be blocked up and it was only with the greatest difficulty that I drew breath at all. Even in the village wheat dust seemed to be everywhere. Only on the roof of the tower did I find any relief. There the air was clean and fresh and soothing. In order to sleep I had to carry my bedding to the highest point and spread it out under the stars. This was fine when I was in South Ridge, but at the sheep pens of Han Family Village there was no escape. There I had to lie down with the rest in the low huts and cough the night away.

The repair crews overcame the breakdowns one by one. Lao Liu, the blacksmith, turned out to be especially able. For half a day I tried in vain to repair a broken sprocket by welding, but the weld always cracked as soon as the cast iron cooled. We had no furnace in which to heat and cool it slowly and so despaired of fixing it at all. That meant one whole machine would be out of action. But Liu said, "Leave it to me. I'm going to make it stronger than the original."

He fashioned an iron ring that was just too small to slip over the extended hub of the sprocket. Then he heated it to a glowing red and the ring expanded just enough to slide on. When it cooled the contracting iron gripped the hub with tremendous pressure. As Liu had promised, it was stronger than the original. Soon the combine was back in action.

And so it was with everything that broke. Liu thought of a way to fix what others had given up on. The tinsmith Ying and the machinist Li were also determined to keep the machines running and often worked far into the night. Just as they were beginning to get things under control two of the tractor drivers made a serious blunder that immobilized one combine completely. They drained the oil from the tractor crankcase, as required by the maintenance manual, and then drove off without refilling it. The bearings immediately seized up and a major overhaul job was necessary before the tractor could run again. Without the tractor the combine was helpless. We tried pulling it with two rubber-tired Fords, which worked on level ground, but not when they came to a slight rise. All attempts to repair the bearings also failed because we couldn't get babbitt metal that would hold. Manager Liu got a specialist to come up from Weihsien, a lathe large enough to turn the bearings was located in Nankung, but it was all useless. Each time the tractor was assembled little pieces of metal kept cracking off the bearing surfaces.

By that time it didn't matter so much. The crews of the other combines were beginning to have some control over their machines. On the fourth day Liu Po-ying's team worked for ten hours without stopping. Their combine rolled around and around humming like a well-oiled watch. Every hour the whistle blew

to announce another tank full of grain and the truck rushed out
to receive it and haul it home. Liu Po-ying had begun to master
the mechanism in his charge. He discovered the key points that
had to be watched, tightened, or adjusted, and assigned crew
members to look after them. Things went so well that he decided
to keep going right through the night. Far away, on the other
side of the flat, the second brigade saw the glow of Po-ying's
light and heard the roar of his motors. They decided not to be
outdone and at midnight two of their machines went back to
work. When I awoke in the morning the expanse of stubble had
vastly expanded. Everyone took heart and the harvesting went
forward at an accelerated pace.

What had apparently been building up to a debacle now suddenly took on all the air of an impending victory. Like a drowning
swimmer who finds all at once that his wild lunges and thrusts
are holding him up, the two brigades pushed vigorously forward
toward the distant shore. I found time to rest a bit and look
around. When a combine stopped in the field I no longer hopped
on the bicycle to find out what was wrong. I waited and usually
the crews themselves solved the problem and started up again.

Night work became the order of the day. Even Big Yang, who
had been plodding around with his usual lethargy, suddenly grew
enthusiastic and took an active hand in setting up two shifts. In
the end he even broke precedent by coming out to ride one of the
machines at night to see what it was like.

Night harvesting turned out to be full of excitement. The great
gray combines sail through the darkness with their running
lights twinkling. Standing on the bridge, holding the wheel that
adjusts the height of the cut, the crew leader feels like the captain
of a ship at sea. The huge hulk beneath him lurches and shifts.
In the stark glare of lights, the great reel walks forward into the
standing wheat. Shorn heads pour onto the canvas conveyor.

Then the long drawn-out whistle wail of a crew in distress
wakes the mechanics from their fitful sleep. They stumble out
into the night, find their bicycles, and go off in the direction of
the sound. A shaft has worked loose, slid sideways in its bearings
and snapped a chain. The shaft itself is bent. There are curses
as the men work on the part in the dark. Their flashlight grows

dim and finally fades away altogether. Someone bruises a finger and swears, "*Ta ma te*—[your mother's]!" under his breath. Finally the shaft is freed from the machine. They race back to the lathe truck with all the broken parts. Blacksmith Liu, who has kept awake by telling stories of factory life to the welder Ying, warms up his forge. Ying gets out his acetylene torch. Machinist Li, who has been studying ideographs under the bright light over the lathe, snaps on an electric drill. In ten minutes they have everything fixed. The mechanics disappear into the night, only to be replaced in the ring of light by another crew carrying a twisted header brace. In the dark they have run into a grave mound at full speed, buried the sickle three feet in the dirt, and crumpled the brace. Here is another hour's work for the repair crew.

In the distance a whistle is wailing again, but this time it is short and sharp. Somewhere out there a grain tank is full. The truck driver, who has been dozing at the wheel, springs into action, starts up his vehicle, slams it into gear, and roars off to receive a new load of wheat.

Chapter 26

People Come First

THE FIRST HARVESTING BRIGADE SETTLED into the sheep huts at Han Family Village as if prepared for a long seige. They laid reed mats over ground that smelled strongly of sheep manure. The huts had no windows, only doorways that let in a shaft of sunlight or moonlight, as the case might be.

The people asleep inside were hot, grease covered, and dripping with sweat, for it was midsummer. The sun had reached the Tropic of Cancer and shone down mercilessly. No one had time to bathe or worry about appearances. Food was dished out in a great cauldron under the open sky. Drivers came in from the field, grabbed bowls from the beam or log where they had laid them down after the last meal, filled them full of hot millet, and ate wherever they could find a place to squat or a brick to sit on. When they were finished they rinsed out their bowls, rolled out their quilts, and lay down to sleep without bothering to undress, for there was no telling when they might be called again.

In the evening when everyone woke up life brightened around the huts. It was then that the crew leaders came together to report to former militiaman Kuo who led the whole combine brigade. Beside an oil lamp placed on a stool Kuo and the brigade record keeper sat cross-legged with their notebooks while the crew leaders told what had happened that day, how many full tanks they had harvested, how much gasoline and oil they had consumed, how many parts had broken or given trouble. They also had a chance to register complaints about supplies, food, and help. This was the time to exchange ideas and experiences.

"You have got to have a system with these machines," said Liu

Po-ying when the others asked him how he kept running so many hours on end. "The three on the combine have to divide their attention and watch different areas for trouble. Otherwise they get in each other's way and no one sees the bolt that is loosening or the chain that is slapping.

"And when you stop for fueling and greasing," he continued, "certain things should be checked and tightened each time, like the bolts and nuts that hold the screens."

"How do you adjust those elevator chains?" asked Li, the intellectual. "We tried them all ways. If you get them too tight they snap as soon as you start. If you get them too loose the links fall part by themselves."

"It's hard to say how we do it," Li Po-ying answered. "But we test it by hand, feel for tension. I can't explain but I can show you. We haven't had any trouble for two days now."

Thus bit by bit we worked out the kinks and helped each other to take the measure of the combines and gradually to master them. Just as the wheat dust covered everything, settled in our ears, made our noses run, stuck eyelids together during sleep, and brought spells of coughing as it lodged in our lungs, so thoughts of gears and engines, shafts and whirling chains pervaded our minds, occupied all the crevices and wrinkles of our brains, and for the time being displaced all other things in our lives.

This was true of the drivers, mechanics, and technicians whose job it was to make the machines work. But Kuo Hu-hsien and Manager Liu were thinking of more important things: how to organize, how to coordinate, how to guarantee the work we were doing. For me the harvest was a problem of broken machinery; for them it was a problem of people and their mutual relations.

I couldn't help remembering Kuo's criticism of my pre-harvest speech. I had said, "The most important thing we have is these machines. If we take care of them well, we can solve all our problems, but if we neglect them they will fail us and everything will be lost."

Kuo came to me later and said: "You have misplaced the emphasis. The most important thing we have is our people. Their knowledge, their welfare, their growth, their cooperation—that is

the first problem and the crucial one. Even if we ruin all th
machines, if our men and women advance in skill and ability t
work together we will be far ahead."

Kuo and Liu assumed from the beginning that the machine
would be mastered sooner or later. What they wanted to stud
was how a harvest should be carried through from beginning t
end, and exchanging information and promoting mechanical ski
was only a part of this. In addition, there was the problem c
coordination between the different sections, the organization c
two shifts, supplies, transport, meals, boiled water. From th
beginning they watched the men, their food, their hours, thei
rest. They encouraged those who needed help and criticize
those who were causing difficulties. In every way they pushe
forward the development of teamwork and cooperation.

One problem in human relations developed into a thorny one
This was the problem posed by the peasants who insisted o
gleaning. A few dishonest persons had already stolen ripe whea
from the farm fields. Now, with their own crops harvested, hor
est peasants came out by the hundreds to glean. Crowds c
women and children followed along behind the combines to pic
up whatever had been missed. In some places this was a lot be
cause the unevenness of the ground and the stunted stalks mad
it impossible for the header to remain low enough everywher
By following after the machines with baskets and bags th
peasants were able to pick up several catties apiece each day

Secretary Li had arranged with the district cadres that no glean
ing would be allowed until the harvest was over. Once th
combines went through and the farm staff, with the help c
soldiers and students, had gleaned all that was considered worth
while, then the word would be passed to the villages to come an
gather the rest. But the villagers didn't wait. They came crowdin
in while the machines were still at work. They got in the way an
were in danger of being run down. The soldiers tried to kee
them back, at least until any given field was finished, but the
had no sooner cleared one field than another would swarm wit
busy gleaners. It was impossible to restrain the crowds.

Some of the cadres, including Secretary Li, were all for callin
in more soldiers, but most were ashamed of the fact that there ha

to be any soldiers at all. It was an admission of failure on their part. They had not properly educated and helped the peasants. As Kuo said, "If the people don't regard the farm as their own, all the soldiers in the world won't protect it. What is needed is more work in the community. We must connect the life of the farm with their lives. We must prove that the machinery is their future as well as ours. And we absolutely have to stop running over their fields and damaging their crops when we cross the flat. How can we expect them to love the farm crops when the farm workers don't love theirs?"

This last was an important point, for much of our equipment was wider than the cart tracks that wound between the peasants' cultivated plots. Whenever we pulled grain drills or combines from one place to another some growing plants were bound to be crushed. There was continued war between cart operators and roadside plot owners anyway. Each peasant tried to protect his own fields by digging deep trenches alongside them. To avoid these trenches, the drivers swerved their carts onto other land and the owners in turn dug pits and embankments to protect their property. In the rainy season the battle reached a climax as mud holes blocked the way and carters detoured through whatever field was handy. Now, to make this even worse we brought in equipment that wouldn't fit on the roads.

The action of the peasants during the harvest brought home to Chiheng Farm that a very important part of their work had failed. The farm had not yet become a community center, no peasants came for help and advice about advanced agricultural methods, high quality seeds, effective insecticides. The farm technicians had few contacts among local people. State farm and peasant each went his separate way. Instead of helping each other, each made things more difficult for the other. Much more effort and hard work were needed to bring about improvement in this sphere.

By noon of the seventh day the harvest was completed. We left the peasants gleaning in the fields, hauled the machines home, and took a day off to celebrate and rest. Everyone walked around with new self-assurance, like veterans of a battle. Hadn't the first mechanized harvest been successfully—even victoriously—completed? What worlds were there to conquer next?

The terrible hours of four to five days past, when it looked as if the whole affair might prove to be a fiasco, were completely forgotten.

But when we totaled up how much wheat we had brought in with eight machines in seven days, everyone sobered up. The whole crop from 12,000 *mou* turned out to be less than 400,000 catties—less than forty catties to the *mou*, less than six bushels to the acre. It was clear then that machinery in itself was not the key to successful large-scale agriculture. We had managed to run the machines without any disasters, but the crop—the crop was nothing less than disastrous. From that day on the work of building state farms in China began to develop on many fronts instead of just one. Agricultural technicians, accountants, and competent managers were seen to be as important as tractor drivers and certainly more important than tractors. Big Yang's dream of solving everything with deep plowing collapsed.

Chapter 27

Pu Chien Tan

RETURNING TO PEKING we rode north all night on the Peking-Hankow line. The coaches were crowded, so the drivers sat in the aisle on their bed rolls. This time they felt more than ever like singing, but as they were scattered through several cars they had to content themselves with telling the other passengers all about the harvest. As the conductor walked through the cars his way was blocked by group after group of suntanned young people engaged in eager conversation with the passengers around them. Soon everyone had heard about the harvest at South Ridge —which was nothing to boast about—and the marvelous machines that accomplished it—which were something else again.

I finally found an empty seat beside three young women from Canton. They insisted in broken Mandarin that the seat was occupied, but when, after half an hour, no one came to claim it, I sat down anyway. An hour later I was startled to discover a fourth girl asleep on the floor with her head and shoulders under one seat and her legs under the opposite seat.

"That's the only way we have found to get any rest," they said. "We've been on the train four days and nights. After a while it is impossible to sit up any longer."

They turned out to be young teachers who had just graduated from the Normal School in Canton. They were on their way to Peking to visit schools and colleges in the capital at the invitation of the Ministry of Education. This was their first trip north and they were very excited.

One of the young women told me that her parents were Christian pastors. "But I have given up Christianity," she said. "We discussed it a lot in school and it seems to me that what

Christians believe is just like a superstition. Morality and ethics surely don't depend on that. There was a lot of preaching in the past, but there was very little morality. Look at China now! Everyone is trying to serve the people. Isn't that real Christian morality? When I get back to Canton I am going to volunteer for a teaching post in Yunnan. I have a friend who has been there for six months already." She smiled as she spoke of it.

"What do your parents think about your views now?" I asked.

"We get along fine," she said. "They think I am wrong but they too have seen that the Communists are working for the good of everyone. They can't deny that. So we argue and argue. Even when I was little I used to wonder about turning the other cheek. Surely the imperialists would walk all over us if we did that. Turn the other cheek, they said, and then sailed a gunboat up the river. It's all so clear to me now.

"This time we are building our own China. We teachers have an enormous job ahead of us. Isn't it wonderful that we are able to come to Peking? I'm so excited I can't sleep anyway. I've never been on a train trip before . . ."

She talked on and on, almost thinking aloud, about religion, about her parents, about New China and her future work.

On the seat to my right a photographer in a blue cadre's uniform listened for a while and then entered into the conversation himself.

"New China! I want to tell you, you don't know what our New China is like until you travel around a bit. There are more things going on than anyone can imagine. Now take me. I've just been down to Chiaotso to the coal mines. Never expected to see such a place. The mines are all mechanized. They have electric trains running underground, and electric lights, and air forced into the shaft by huge fans. But the best thing is the new homes. There are tens of thousands of workers all moving into new rooms. We do not have dwellings like that, even in Peking. They all have running water and lights. There is a hospital and a theater and schools for the children, just like a regular city. And that's just the beginning. There is lots more coal there. The mines will be much bigger in the future.

"The newspaper sent me there. We had heard it was remark-

able, but you can't imagine until you've seen it. When the lights
go on at night you can see a whole valley of lights.

"And now here is a fellow who has been harvesting wheat with
combines. Who ever heard of that before? Yet we're already
doing it. I tell you you have to travel around a bit to find out what
is going on. New China is *pu chien tan* [not so simple]."

In the small hours just before dawn we all finally dozed off.
The train carried us swiftly northward through Tinghsien and
Paoting toward the great iron bridge over the Yungting River.
Fengtai's long railroad yards flashed past and then we came to
Peking's Great Wall. In the golden glow of morning the city
looked peaceful, quiet, and clean. The blue roof of the Temple
of Heaven showed itself over the battlements of the outer city as
we glided through mile after mile of carefully tended truck
gardens glistening with a bright sheen of dew.

As the train entered the station, the pulse of life quickened. The
platform was jammed with people loaded down with bundles
and parcels. Inside the waiting rooms and out on the sidewalk,
others sat on their baggage and waited in orderly lines for the
trains that would take them to the far corners of the country,
to Shanghai, Canton, Sian, Manchouli, and Paotou. Peking had
truly become the hub of the nation. No matter what your rail
destination, there was always a train that went there direct from
Peking.

The public security officer at the platform gate scanned the
moving line for foreigners. When he saw me he tried to look
efficient and aloof. But there was a softness in his eye. Finally,
after examining my pass, he said, "Well, Lao Han, what have
you been up to this time?"

"This time it was the wheat harvest. A mechanized harvest.
China's first."

"How did it go?"

"Swell, it went just swell."

"Wonderful," said the officer. "That's wonderful. Our New
China, it's *pu chien tan*."

Postscript, 1970

THE GREAT PROLETARIAN CULTURAL REVOLUTION has ushered in a whole new stage in the Chinese Revolution. It has also opened the previous stages to searching mass examination in a vast effort to expose and ultimately overthrow the authority and influence of the bourgeoisie in Chinese society, in the Chinese Revolution, and in the Chinese Communist Party. No serious study of any aspect of the Revolution can escape re-examination now that the "struggle between two lines"—the struggle between the socialist road and the capitalist road, between the proletarian headquarters and the bourgeois headquarters, between the Marxist-Leninists and the revisionists—has been so thoroughly revealed.

Iron Oxen is no exception. This book was written as a recollected journal of the author's life and work in China during 1949 and 1950. It describes the first steps in the creation of a system of mechanized state farms as a critical turning point in the Chinese Revolution—the end of imperialist-Kuomintang rule and the establishment of the People's Republic of China. It reflects the consciousness of these events which the author held at the time.

The year 1949 was one of great political complexity. With hindsight it is now evident that what I, together with the cadres and students of the tractor program, lived through in those days was not simply victory for the new-democratic class coalition which Mao Tse-tung had skillfully forged and led for so many years, but the end of the new-democratic stage and the beginning of the socialist stage in the Chinese Revolution as a whole.

What I was conscious of at the time, however, was not a great transition from one stage of the Revolution to another, but a

straight-line development of new democracy, its culmination, so
to speak: the smashing of foreign intervention on the mainland
the destruction of feudal landholdings; the creation of a mixed
economy of state-owned, cooperative, and private enterprises ex
isting and expanding side by side; a continuing alliance of four
classes—workers, peasants, petty bourgeoisie, and national bour
geoisie—led by the working class through its party, the Com
munist Party; a Common Program which stood as the manifesto
of this class alliance and served as the framework for judging
all programs and all actions.

How long would this stage last? Quite a long time I assumed
Many statements by Chinese Communist leaders, including some
by Mao Tse-tung himself, pointed to decades rather than year

In regard to socialism, I understood that this was the next goal
I also understood that the state farm program, in which I worked
was a part of the socialist component of new democracy, that
this component was decisive in the economy (it included all the
heavy industry and transport expropriated from old China's four
leading families), and that it represented China's future. What
I did not grasp was that the socialist revolution had already
begun, that just as new democracy defined the content and basic
class relationships of the revolution to overthrow Chinese feu
dalism and foreign imperialism in the period from 1919 to 1949, so
socialism defined the content and basic class relations of the rev
olution to collectivize agriculture and to restrict, limit, and finally
to expropriate the bourgeoisie after 1949. Before that year
the society around us had been feudal, comprador, imperialist
dominated. The Liberated Areas where I had worked for three
years had been islands of new democracy thrust up in a feudal
comprador-imperialist sea. After 1949 the mixed economy, poli
tics, and culture of new democracy soon dominated the whole
nation and what constituted the new was no longer new de
mocracy but the islands of socialism, among them state farms
that thrust up in the new-democratic sea. Just as the struggle to
make new democracy universal had constituted the central task
of the old stage, so the struggle to make socialism universal con
stituted the central task of the new. And this was true even

though in 1949 millions of peasants in south and west China had
not yet settled accounts with their landlords or expropriated their
property. Land reform, which loomed so large and absorbed so
much energy in those days, was still in the nature of a left-over
task. The center of the agrarian struggle had already shifted
elsewhere—to the development of mutual-aid groups, land pooling
cooperatives, and collective farms by peasants who had *fanshened*.

The victory of 1949, it is now clear, had by no means ended
class struggle in China; it had only transformed it. Having re-
solved, at least internally, the great contradiction between the
Chinese people—primarily the workers and peasants—on the one
hand, and the feudal landlords, the bureaucratic-comprador
capitalists of the Kuomintang, and the American imperialists on
the other, victory shifted a previously subordinate contradiction
to the center of the stage. This was the contradiction between
the working class and the bourgeoisie, including the national
bourgeoisie which was still an important component of new-
democratic society. The red thread of this new conflict ran
through every sphere of Chinese life and every aspect of China's
reconstruction. It took the form of a contest between the "capitalist road" and the "socialist road" in farming, in industry, in trade,
in education, in culture, and in politics.

In the countryside the shock forces of this struggle were the
former poor and lower-middle peasants who saw no future in
small-plot farming and moved to collectivize, step by step, in
the face of protracted opposition from the more prosperous
peasants. In the cities the shock force was made up of rank-and-
file workers who struggled to give real socialist content to
state-owned industry and trade while riding herd on the private
sector until it too could be transformed. Inside the Communist
Party itself both these roads had adherents: Mao Tse-tung re-
peatedly took the initiative for socialist transformation in all
spheres, and Liu Shao-chi consistently applied the brakes with
policies that objectively fostered a private economy and an ex-
tension of bourgeois, "humanistic" culture.

It is important to raise all this because there is a tendency in
Iron Oxen to assume that the great revolutionary storm was over,

that the working class had been victorious in a final sense, and
that the crucial job at hand was economic, technical, and administrative—rather than political. The reader could well be lulled into
thinking that with the Kuomintang smashed and the landlords
expropriated, the main problem in the countryside was modernization—that favorite word of American academia—a modernization
that included, as an important component, mechanization. If only
the young peasants and workers who were the students at
Chiheng could master internal combustion engines and combine
harvesters, if only their counterparts in the agricultural schools
could master soil science, insecticides, and plant growth, the
future of socialism in China was assured.

All such assumptions have been proven false by the development of socialist revolution not only in China but throughout
the world. Modernization without conscious and protracted class
struggle to ensure that state power remains in working-class
hands and that the superstructure of society, as well as its base,
is transformed, can and has led back to capitalism. False though
these assumptions were and are, they were held to a greater or
lesser degree by many revolutionary Chinese and they influenced
the writing of this book. The reappraisal of history that has
accompanied the Cultural Revolution demonstrates that this was
no accident. A lot of confused thinking about the problems of
the transition to socialism existed in China after 1949, not only
among the rank-and-file but within the Communist Party and its
leading bodies. Behind the confusion lay a clear-cut conflict
between factions led respectively by Mao Tse-tung and Liu
Shao-chi. This conflict crystalized, in part, around the theory
of the productive forces, which is an essential component of
Marxism.

Liu Shao-chi and his followers held to a version of this
theory reminiscent of Bukharin: since the productive forces—the
land, resources, machinery, techniques, and labor power—available
to any society determine its contours and development, and since
socialism is a historical stage requiring a high degree of division
of labor, industrialization, technical knowledge, and scientific
research, it is impossible to develop socialist forms of agricultural

organization in the Chinese countryside until industry is able to provide tractors, fertilizers, and insecticides and the schools can provide a large cadre of technicians, scientists, etc. China's peasants should be content to take the rich peasant road (the capitalist road) until the urban sector can provide a sound technical base for a switch to socialism.

Mao and his followers, on the other hand, held to a different version of the theory: they recognized not only the importance of the productive forces but also the vital impact of the relations of production—class structure, forms of ownership, forms of organization among producers—on these forces. Old, outmoded relations of production can inhibit, even destroy, productive forces; new, revolutionary productive relations can develop and even create new productive forces. Furthermore, Mao recognized the importance of the superstructure—that is, ideology, education, the arts, law, and all the institutions outside the sphere of production—on productive relations and on the forces of production. It was Mao's view that, under the conditions existing in China, only a fundamental shift in productive relations in the countryside—a shift from individual to collective agriculture—could lay the groundwork for an all-round development of productive forces, provide a market for China's growing industries, supply these industries with raw materials, and make possible the introduction of tillage machinery, electricity, pumps, and modern techniques of all kinds into the Chinese countryside.

"Socialist industrialization cannot be carried out in isolation, separately from agricultural cooperation," Mao wrote in 1955.

> Heavy industry, the most important branch of socialist industrialization, produces tractors and other agricultural machinery, chemical fertilizers, modern means of transport, oil, electric power, and other items for the needs of agriculture, but it is only on the basis of a large-scale cooperative agriculture that it is possible to use all these, or use them on an extensive scale. We are now carrying out a revolution not only in the social system, changing from private ownership to public ownership, but also in technology, changing from handicraft production to mass production with up-to-date

machinery; and these two revolutions are interlinked. In agriculture, under the conditions prevailing in our country, cooperation must precede the use of big machinery.

The theory advocated by Liu Shao-chi was an expression of economic determinism and served as an excuse for a policy of laissez-faire that could only have resulted in a rapid class differentiation in the Chinese countryside, stagnant productive forces, and a stagnant market leading to stagnation in industry and an eventual return to semi-feudal, semi-colonial status for China. The theory advocated by Mao Tse-tung was an expression of dialectical materialism. Applied in practice, it resulted in the socialization of Chinese agriculture through class struggle and a relatively rapid development of rural productive forces. These latter were greatly stimulated by the large scale of the new agriculture, its potential for rational land and resource use, its capacity for capital accumulation, and its mobilization of all labor power—off-season, part-time, handicapped, etc.—for sustained production work.

Looking backward, it is not hard to see that many of us in the tractor program were greatly influenced by Liu's theories. Cadres and workers on state farms thought of themselves as the vanguard of Chinese agriculture, who, having mastered modern technique, would lead the entire peasantry down the high road of large-scale socialist agriculture sometime in the future. We were quite aware that we were part of the socialist sector of the new-democratic society, and since this sector was decisive in the economy as a whole we naturally thought of ourselves as the decisive factor in the transformation of agriculture, the advance guard. Actually, it now seems clear that the advance guard of socialism in Chinese agriculture in 1949 consisted of poor peasants like those in Wang Kuo-fan's pauper's co-op, which was made up of twenty-three families who, beside their land, owned only a three-quarter share in a donkey. "But relying on their own efforts, in three years' time its members accumulated a large quantity of the means of production," wrote Mao Tse-tung.

They "got it from the mountains," they explained. Some of the people visiting the cooperative were moved to tears when they learned what this meant. Our entire nation, we feel, should pattern itself after this co-op. In a few decades, why can't 600 million "paupers," by their own efforts, create a socialist country, rich and strong?*

Hundreds of millions of poor and lower-middle peasants took up Mao's challenge in the middle fifties. They consciously chose the socialist road and it was this decisive battle against the capitalist road which they fought and won which gave importance to the mechanization pioneered by the state farms.

Any reader who is unaware of this could be misled by the sub-title and text of *Iron Oxen*. The book *is* a documentary of revolution in Chinese farming, but it describes only one aspect of that revolution, an aspect that was important to the revolution as a whole because the peasants of China had to master modern technique if they were ever to take command of nature and fashion a flourishing socialist society, but an aspect, nevertheless, which was subordinate to the struggle to take the socialist road that occurred in hundreds of thousands of obscure villages and hamlets throughout the length and breadth of the land. The true sequel to *Fanshen* lies not in *Iron Oxen* alone, but in books like *The Builders* by Liu Ching, where the first steps toward mutual aid in an isolated village after land reform are described; in books like *Socialist Upsurge in China's Countryside*, where the experiences of peasant activists and rural cadres in building producers' cooperatives in various parts of China prior to 1956 are collected between two covers and edited by Mao Tse-tung; in books like *Great Changes in a Mountain Village* by Chou Li-po, and *The Rise of the People's Communes in China* by Anna Louise Strong.

Readers of *Iron Oxen* may of course discern that the struggle between the capitalist road and the socialist road after land reform took place not only in the private sector of China's economy —that is, among millions of small-holding peasants—but also in the socialist sector, in the Tractor Training Center at Chihsien (later at Double Bridge), in the State Farm Management Bureau, and on the various state farms that were created as full-blown socialist enterprises in 1949 and after.

---

* *Socialist Upsurge in China's Countryside*, 1956.

One of the great lessons to be learned from the Cultural Revolution is that the establishment of socialist forms is not the end of the revolution on the economic or any other front.

> After the realization of agricultural cooperation . . . the struggle between the consolidation of the socialist system of collective ownership and the attempt to sabotage it remains an outstanding question. . . . The proletariat and the former poor and middle peasants must use the tremendous power of the dictatorship of the proletariat to consolidate and develop the socialist system of collective ownership so as to take the road of common prosperity.

So reads an article entitled "Struggle in China's Countryside Between the Two Roads," which appeared in *Hsinhua News* on December 4, 1967. If this was true in cooperatives and communes established by peasants, it was no less true on state farms established directly by government cadres.

The most striking example of this struggle in the text is the story of Manager Li, from Chiheng Farm, who used capital and transport alloted for the development of crop production to build a small trading empire and thus convert a socialist farm into an institution for capitalist profit. Other examples of this struggle appear in the meetings for self-and-mutual criticism, where individualism, hedonism, and go-it-alone philosophies clash with the strivings for mutual aid, collective consciousness, and unconditional service to the people that are fundamental to socialism. It is also revealed in the contrast between my view that "These machines are the most important thing we have," and Kuo Hu-hsien's view that "It is the people who drive them who are most important." Where capitalism puts machines and technique in first place, socialism puts man and man's political consciousness first.

As the years passed, the struggle between the two roads in state farm construction became sharper and more clearly defined. It showed up most strikingly in the battle between the foreign-trained experts assigned as teachers in the tractor school and Director Li (not to be confused with Manager Li) over the form and content of the education provided. The experts wanted to build a school and later a college that resembled the institutions of higher learning that they themselves had known in China and the West. They wanted strict selection by examination both for admission to the school and advancement from class to class. They wanted students graded according to knowledge already mastered, with advanced courses for advanced students and the culling out of the slow and the backward. They resisted and resented "practice"—the periodic emptying of the school as faculty and students alike went down to the farms to plow, plant, and harvest. They felt that their own careers had been aborted by exile to a mere "trade school." By rights they should be heading departments and research facilities at great universities. They also demanded the privileges and bonuses, salaries and distinctions, that traditionally went with such high-level posts.

Director Li, who had worker and peasant support, had an entirely different approach to education. He was for taking in all students assigned to the school, regardless of level, and for advancing them all, by whatever means necessary, toward the level of master technician without demoting anyone or culling anyone out. He was for mixed classes of culturally advanced and culturally backward students. He was for mutual aid between students of all levels and for mutual aid between those with practical experience and those with theoretical training. He was for the most intimate combination of theory and practice, with the school actually operating a farm and with staff and students periodically joining production elsewhere. And he was for research and development in the field arising out of real work and carried on by those engaged in the actual process of production. He also insisted on political mobilization, the fostering of proletarian consciousness, as a prerequisite for building a new socialist world.

The foreign-trained experts wanted to develop a technical elite on the Western model. Consciously or unconsciously, they were advocates of the capitalist road. Li Chih was for the all-round liberation and development of the talents of the working people of the state farms. He was, I feel, advocating the socialist road.

But though this struggle between the two roads was actually taking place and I took part in it on Director Li's side, I was not conscious of it's being part of a basic class struggle developing in a new period of socialist revolution. Thus the text of *Iron Oxen* expresses a certain naiveté, a certain innocence, that more accurately reflects the exuberant, euphoric state of mind quite common in the Liberated China of those years, than the actual situation that was unfolding—a situation of sharp class struggle between proletariat and bourgeoisie in every field from crop production to playwriting, and at every level from peasant plot to state council.

In the context of the times, this state of mind can perhaps be excused. The struggle of the Chinese people against feudalism and imperialism had been long and bitter. The chronic violence, misery, and oppression of the old society had pushed people to the limits of endurance and they had paid a heavy price in mounting the revolt that smashed it. The victory of 1949 brought in its wake a tremendous liberation not only of body but of mind and spirit. As Wordsworth said of the first days of the French Revolution:

> Bliss it was in that dawn to be alive,
> But to be young was very heaven!

It would be inexcusable, however, in 1970, not to point out the limits of this earlier understanding, and to arm the reader with a more complete insight into the actual development of the Chinese Revolution. For it is exactly this innocence, this naiveté in regard to class struggle, on which the bourgeoisie counted and still counts to allay the people's vigilance and provide the conditions for a renewed assault on socialism and proletarian state power after it has been established anywhere. Contrary to the line put forward in 1956 by Liu Shao-chi that "In China, the question of which wins out, socialism or capitalism, is already solved," Mao Tse-tung specifically pointed out:

> The question of which will win out, socialism or capitalism, is still not really settled. . . .
>
> The class struggle between the proletariat and the bourgeoisie, the class struggle between the different political

forces, and the class struggle in the ideological field between the proletariat and the bourgeoisie will continue to be long and tortuous and at times will even become very acute. . . .

After the enemies with guns have been wiped out, there will still be the enemies without guns; they are bound to struggle desperately against us, and we must never regard these enemies lightly. If we do not now raise and understand the problem in this way, we shall commit the gravest mistakes.

In a world where struggle against imperialists is inextricably linked with struggle against revisionists—bourgeoisie, new and old, masquerading as socialists—it would indeed be a grave mistake to gloss over or obscure this problem. No revolution can possibly succeed without a clear grasp of the new ways in which the old ruling classes distort and misdirect revolutionary movements from within before their own state power is smashed and then once again seize power from the working class after working-class states are created.

The great teacher of these lessons is Mao Tse-tung. The Cultural Revolution he is leading in China is a practical demonstration of how the working class can and must defend its hard-won state power, defeat revisionism, and carry the socialist revolution through to the end.

WILLIAM HINTON first visited China in 1937, where, after working six months as a newspaper reporter in Japan, he traveled across Manchuria on his way home. He saw much more of the country in 1945 while working as a propaganda analyst for the United States Office of War Information in several Chinese cities. He returned to China in 1947 and stayed until 1953. His book *Fanshen* is also available in Vintage Books. Hinton now runs a farm in Pennsylvania, where he lives with his wife and three children.

VINTAGE POLITICAL SCIENCE
AND SOCIAL CRITICISM

V-428   ABDEL-MALEK, ANOUAR   *Egypt: Military Society*
V-625   ACKLAND, LEN AND SAM BROWN   *Why Are We Still in Vietnam?*
V-340   ADAMS, RUTH (ed.)   *Contemporary China*
V-196   ADAMS, RICHARD N.   *Social Change in Latin America Today*
V-568   ALINSKY, SAUL D.   *Reveille for Radicals*
V-365   ALPEROVITZ, GAR   *Atomic Diplomacy*
V-503   ALTHUSSER, LOUIS   *For Marx*
V-286   ARIES, PHILIPPE   *Centuries of Childhood*
V-511   BAILEY, STEPHEN K.   *Congress Makes a Law*
V-604   BAILYN, BERNARD   *Origins of American Politics*
V-334   BALTZELL, E. DIGBY   *The Protestant Establishment*
V-335   BANFIELD, E. G. AND J. Q. WILSON   *City Politics*
V-674   BARBIANA, SCHOOL OF   *Letter to a Teacher*
V-198   BARDOLPH, RICHARD   *The Negro Vanguard*
V-185   BARNETT, A. DOAK   *Communist China and Asia*
V-270   BAZELON, DAVID   *The Paper Economy*
V-60   BECKER, CARL L.   *The Declaration of Independence*
V-563   BEER, SAMUEL H.   *British Politics in the Collectivist Age*
V-199   BERMAN, H. J. (ed.)   *Talks on American Law*
V-211   BINKLEY, WILFRED E.   *President and Congress*
V-81   BLAUSTEIN, ARTHUR I. AND ROGER R. WOOCK (eds.)   *Man Against Poverty*
V-508   BODE, BOYD H.   *Modern Educational Theories*
V-513   BOORSTIN, DANIEL J.   *The Americans: The Colonial Experience*
V-358   BOORSTIN, DANIEL J.   *The Americans: The National Experience*
V-621   BOORSTIN, DANIEL J.   *The Decline of Radicalism: Reflections on America Today*
V414   BOTTOMORE, T. B.   *Classes in Modern Society*
V-44   BRINTON, CRANE   *The Anatomy of Revolution*
V-625   BROWN, SAM AND LEN ACKLAND   *Why Are We Still in Vietnam*
V-234   BRUNER, JEROME   *The Process of Education*
V-590   BULLETIN OF ATOMIC SCIENTISTS   *China after the Cultural Revolution*
V-578   BUNZEL, JOHN H.   *Anti-Politics in America*
V-549   BURNIER, MICHEL-ANTOINE   *Choice of Action*
V-684   CALVERT, GREG AND CAROL   *The New Left and the New Capitalism*
V-30   CAMUS, ALBERT   *The Rebel*
V-33   CARMICHAEL, STOKELY AND CHARLES HAMILTON   *Black Power*
V-664   CARMICHAEL, STOKELY   *Stokely Speaks*
V-98   CASH, W. J.   *The Mind of the South*
V-556   CASTRO, JOSUE DE   *Death in the Northeast*
V-272   CATER, DOUGLASS   *The Fourth Branch of Government*
V-290   CATER, DOUGLASS   *Power in Washington*
V-551   CHEVIGNY, PAUL   *Police Power*
V-555   CHOMSKY, NOAM   *American Power and the New Mandarins*
V-640   CHOMSKY, NOAM   *At War With Asia*
V-554   CONNERY, ROBERT H. (ed.)   *Urban Riots: Violence and Social Change*

V-420   CORNUELLE, RICHARD C.   *Reclaiming the American Dream*

V-538   COX COMMISSION   *Crisis at Columbia*

V-311   CREMIN, LAWRENCE A.   *The Genius of American Education*

V-519   CREMIN, LAWRENCE A.   *The Transformation of the School*

V-734   DANIELS, R. V.   *A Documentary History of Communism*

V-237   DANIELS, R. V.   *The Nature of Communism*

V-638   DENNISON, GEORGE   *The Lives of Children*

V-746   DEUTSCHER, ISAAC   *The Prophet Armed*

V-747   DEUTSCHER, ISAAC   *The Prophet Unarmed*

V-748   DEUTSCHER, ISAAC   *The Prophet Outcast*

V-617   DEVLIN, BERNADETTE   *The Price of My Soul*

V-671   DOMHOFF, G. WILLIAM   *The Higher Circles*

V-603   DOUGLAS, WILLIAM O.   *Points of Rebellion*

V-645   DOUGLAS, WILLIAM O.   *International Dissent*

V-585   EAKINS, DAVID AND JAMES WEINSTEIN (eds.)   *For a New America*

V-390   ELLUL, JACQUES   *The Technological Society*

V-379   EMERSON, T. I.   *Toward a General Theory of the First Amendment*

V-47   EPSTEIN, B. R. AND A. FORSTER   *The Radical Right: Report on the John Birch Society and Its Allies*

V-692   EPSTEIN, JASON   *The Great Conspiracy Trial*

V-661   FALK, RICHARD A., GABRIEL KOLKO, AND ROBERT JAY LIFTON   *Crimes of War: After Songmy*

V-442   FALL, BERNARD B.   *Hell in a Very Small Place: The Siege of Dien Bien Phu*

V-423   FINN, JAMES   *Protest: Pacifism and Politics*

V-667   FINN, JAMES   *Conscience and Command*

V-225   FISCHER, LOUIS (ed.)   *The Essential Gandhi*

V-424   FOREIGN POLICY ASSOCIATION, EDITORS OF   *A Cartoon History of United States Foreign Policy Since World War I*

V-413   FRANK, JEROME D.   *Sanity and Survival*

V-382   FRANKLIN, JOHN HOPE AND ISIDORE STARR (eds.)   *The Negro in 20th Century America*

V-224   FREYRE, GILBERTO   *New World in the Tropics*

V-368   FRIEDENBERG, EDGAR Z.   *Coming of Age in America*

V-662   FREIDMAN, EDWARD AND MARK SELDEN (eds.)   *America's Asia: Dissenting Essays in Asian Studies*

V-378   FULBRIGHT, J. WILLIAM   *The Arrogance of Power*

V-264   FULBRIGHT, J. WILLIAM   *Old Myths and New Realties and other Commentaries*

V-354   FULBRIGHT, J. WILLIAM (intro.)   *The Vietnam Hearings*

V-688   FULBRIGHT, J. WILLIAM   *The Pentagon Propaganda Machine*

V-461   GARAUDY, ROGER   *From Anathema to Dialogue*

V-561   GALIN, SAUL AND PETER SPIELBERG (eds.)   *Reference Books: How to Select and Use Them*

V-475   GAY, PETER   *The Enlightenment: The Rise of Modern Paganism*

V-277   GAY, PETER   *Voltaire's Politics*

V-668   GERASSI, JOHN   *Revolutionary Priest: The Complete Writings and Messages of Camillo Torres*

V-657   GETTLEMAN, MARVIN E. AND DAVID MERMELSTEIN (eds.)   *The Failure of American Liberalism*

V-451   GETTLEMAN, MARVIN E. AND SUSAN, AND LAWRENCE AND CAROL KAPLAN   *Conflict in Indochina: A Reader on the Widening War in Laos and Cambodia*

V-174   GOODMAN, PAUL AND PERCIVAL   *Communitas*

V-325   GOODMAN, PAUL   *Compulsory Mis-education and The Community of Scholars*

V-32   GOODMAN, PAUL   *Growing Up Absurd*

V-417   GOODMAN, PAUL   *People or Personnel* and *Like a Conquered Province*

V-247   GOODMAN, PAUL   *Utopian Essays and Practical Proposals*

V-606   GORO, HERB   *The Block*

V-633   GREEN, PHILIP AND SANFORD LEVINSON (eds.)   *Power and Community: Dissenting Essays in Political Science*

V-457   GREENE, FELIX   *The Enemy: Some Notes on the Nature of Contemporary Imperialism*

V-618   GREENSTONE, J. DAVID   *Labor in American Politics*

V-430   GUEVERA, CHE   *Guerrilla Warfare*

V-685   HAMSIK, DUSAN   *Writers Against Rulers*

V-605   HARRIS, CHARLES F. AND JOHN A. WILLIAMS (eds.)   *Amistad 1*

V-660   HARRIS, CHARLES F. AND JOHN A. WILLIAMS (eds.)   *Amistad 2*

V-427   HAYDEN, TOM   *Rebellion in Newark*

V-453   HEALTH PAC   *The American Health Empire*

V-635   HEILBRONER, ROBERT L.   *Between Capitalism and Socialism*

V-404   HELLER, WALTER (ed.)   *Perspectives on Economic Growth*

V-450   HERSH, SEYMOUR M.   *My Lai 4*

V-283   HENRY, JULES   *Culture Against Man*

V-644   HESS, KARL AND THOMAS REEVES   *The End of the Draft*

V-465   HINTON, WILLIAM   *Fanshen: A Documentary of Revolution in a Chinese Village*

V-576   HOFFMAN, ABBIE   *Woodstock Nation*

V-95   HOFSTADTER, RICHARD   *The Age of Reform: From Bryan to F.D.R.*

V-9   HOFSTADTER, RICHARD   *The American Political Tradition*

V-317   HOFSTADTER, RICHARD   *Anti-Intellectualism in American Life*

V-385   HOFSTADTER, RICHARD   *Paranoid Style in American Politics and other Essays*

V-686   HOFSTADTER, RICHARD AND MICHAEL WALLACE (eds.)   *American Violence, A Documentary History*

V-429   HOROWITZ, DE CASTRO, AND GERASSI (eds.)   *Latin American Radicalism*

V-666   HOWE, LOUISE KAPP (ed.)   *The White Majority: Between Poverty and Affluence*

V-630   HOROWITZ, DAVID   *Empire and Revolution*

V-201   HUGHES, H. STUART   *Consciousness and Society*

V-514   HUNTINGTON, SAMUEL F.   *The Soldier and the State*

V-241   JACOBS, JANE   *Death & Life of Great American Cities*

V-584   JACOBS, JANE   *The Economy of Cities*

V-433   JACOBS, PAUL   *Prelude to Riot*

V-332   JACOBS, PAUL AND SAUL LANDAU (eds.)   *The New Radicals*

V-459   JACOBS, PAUL AND SAUL LANDAU, WITH EVE PELL   *To Serve the Devil: Natives and Slaves*, Vol. I

V-460   JACOBS, PAUL AND SAUL LANDAU, WITH EVE PELL   *To Serve the Devil: Colonials & Sojourners*, Vol. II

V-456   JONES, ITA   *The Grubbag*

V-451   KAPLAN, LAWRENCE AND CAROL, MARVIN E. AND SUSAN GETTLEMAN   *Conflict in Indochina: A Reader on the Widening War in Laos and Cambodia*

V-369   KAUFMANN, WALTER (trans.)   *The Birth of Tragedy* and *The Case of Wagner*

V-401	KAUFMANN, WALTER (trans.) *On the Genealogy of Morals* and *Ecce Homo*
V-337	KAUFMANN, WALTER (trans.) *Beyond Good and Evil*
V-482	KELSO, LOUIS O. AND PATRICIA HETTER *Two-Factor Theory: The Economics of Reality*
V-470	KEY, V. O. JR. *The Responsible Electorate*
V-510	KEY, V. O. *Southern Politics*
V-341	KIMBALL & MCCLELLAN *Education and the New America*
V-582	KIRSHBAUM, LAURENCE AND ROGER RAPOPORT *Is the Library Burning?*
V-631	KOLKO, GABRIEL *Politics of War*
V-661	KOLKO, GABRIEL, RICHARD A. FALK AND ROBERT JAY LIFTON (eds.) *Crimes of War: After Songmy*
V-361	KOMAROVSKY, MIRRA *Blue-Collar Marriage*
V-675	KOVEL, JOVEL *White Racism*
V-215	LACOUTURE, JEAN *Ho Chi Minh*
V-459	LANDAU, SAUL, PAUL JACOBS, WITH EVE PELL *To Serve the Devil: Natives and Slaves*, Vol. I
V-460	LANDAU, SAUL, PAUL JACOBS, WITH EVE PELL *To Serve the Devil: Colonials & Sojourners*, Vol. II
V-367	LASCH, CHRISTOPHER *The New Radicalism in America*
V-560	LASCH, CHRISTOPHER *The Agony of the American Left*
V-399	LASKI, HAROLD J. (ed.) *Harold J. Laski on the Communist Manifesto*
V-426	LEKACHMAN, ROBERT *The Age of Keynes*
V-638	LEVINSON, SANFORD AND PHILIP GREEN (eds.) *Power and Community: Dissenting Essays in Political Science*
V-280	LEWIS, OSCAR *The Children of Sánchez*
V-421	LEWIS, OSCAR *La Vida*
V-370	LEWIS, OSCAR *Pedro Martínez*
V-284	LEWIS, OSCAR *Village Life in Northern India*
V-634	LEWIS, OSCAR *A Death in the Sánchez Family*
V-637	LIBARLE, MARC AND TOM SELIGSON (eds.) *The High School Revolutionaries*
V-392	LICHTHEIM, GEORGE *The Concept of Ideology and Other Essays*
V-474	LIFTON, ROBERT JAY *Revolutionary Immortality*
V-661	LIFTON, ROBERT JAY, RICHARD A. FALK AND GABRIEL KOLKO (eds.) *Crimes of War: After Songmy*
V-690	LIFTON, ROBERT JAY *History and Human Survival*
V-384	LINDESMITH, ALFRED *The Addict and The Law*
V-533	LOCKWOOD, LEE *Castro's Cuba, Cuba's Fidel*
V-469	LOWE, JEANNE R. *Cities in a Race with Time*
V-659	LURIE, ELLEN *How to Change the Schools*
V-193	MALRAUX, ANDRE *Temptation of the West*
V-480	MARCUSE, HERBERT *Soviet Marxism*
V-502	MATTHEWS, DONALD R. *U. S. Senators and Their World*
V-552	MAYER, ARNO J. *Politics and Diplomacy of Peacemaking*
V-577	MAYER, ARNO J. *Political Origins of the New Diplomacy, 1917-1918*
V-575	MCCARTHY, RICHARD D. *The Ultimate Folly*
V-619	MCCONNELL, GRANT *Private Power and American Democracy*
V-386	MCPHERSON, JAMES *The Negro's Civil War*
V-657	MERMELSTEIN, DAVID AND MARVIN E. GETTLEMAN (eds.) *The Failure of American Liberalism*
V-273	MICHAEL, DONALD N. *The Next Generation*

V-19   MILOSZ, CZESLAW   *The Captive Mind*
V-669   MINTZ, ALAN L. AND JAMES A. SLEEPER   *The New Jews*
V-615   MITFORD, JESSICA   *The Trial of Dr. Spock*
V-316   MOORE, WILBERT E.   *The Conduct of the Corporation*
V-539   MORGAN, ROBIN (ed.)   *Sisterhood is Powerful*
V-251   MORGENTHAU, HANS J.   *The Purpose of American Politics*
V-57   MURCHLAND, BERNARD (ed.)   *The Meaning of the Death of God*
V-274   MYRDAL, GUNNAR   *Challenge to Affluence*
V-573   MYRDAL, GUNNAR   *An Approach to the Asian Drama*
V-687   NEVILLE, RICHARD   *Play Power*
V-377   NIETZSCHE, FRIEDRICH   *Beyond Good and Evil*
V-369   NIETZSCHE, FRIEDRICH   *The Birth of Tragedy* and *The Case of Wagner*
V-401   NIETZSCHE, FRIEDRICH   *On the Genealogy of Morals* and *Ecce Homo*
V-689   OBSERVER, AN   *Message from Moscow*
V-642   O'GORMAN, NED   *Prophetic Voices*
V-583   ORTIZ, FERNANDO   *Cuban Counterpoint: Tobacco and Sugar*
V-285   PARKES, HENRY B.   *Gods and Men*
V-624   PARKINSON, G. H. R.   *Georg Lukacs: The Man, His Work, and His Ideas*
V-128   PLATO   *The Republic*
V-648   RADOSH, RONALD   *American Labor and U. S. Foreign Policy*
V-582   RAPOPORT, ROGER AND LAURENCE KIRSHBAUM   *Is the Library Burning?*
V-309   RASKIN, MARCUS and BERNARD FALL (eds.)   *The Viet-Nam Reader*
V-719   REED, JOHN   *Ten Days That Shook the World*
V-644   REEVES, THOMAS and KARL HESS   *The End of the Draft*
V-192   REISCHAUER, EDWIN O.   *Beyond Vietnam: The United States and Asia*
V-548   RESTON, JAMES   *Sketches in the Sand*
V-622   ROAZEN, PAUL   *Freud: Political and Social Thought*
V-534   ROGERS, DAVID   *110 Livingston Street*
V-559   ROSE, TOM (ed.)   *Violence in America*
V-212   ROSSITER, CLINTON   *Conservatism in America*
V-472   ROSZAK, THEODORE (ed.)   *The Dissenting Academy*
V-288   RUDOLPH, FREDERICK   *The American College and University*
V-408   SAMPSON, RONALD V.   *The Psychology of Power*
V-431   SCHELL, JONATHAN   *The Village of Ben Suc*
V-403   SCHRIEBER, DANIEL (ed.)   *Profile of the School Dropout*
V-375   SCHURMANN, F. and O. SCHELL (eds)   *The China Reader: Imperial China, I*
V-376   SCHURMANN, F. and O. SCHELL (eds.)   *The China Reader: Republican China, II*
V-377   SCHURMANN, F. and O. SCHELL (eds.)   *The China Reader: Communist China, III*
V-394   SEABURY, PAUL   *Power, Freedom and Diplomacy*
V-649   SEALE, BOBBY   *Seize the Time*
V-662   SELDEN, MARK AND EDWARD FRIEDMAN (eds.)   *America's Asia: Dissenting Essays in Asian Studies*
V-637   SELIGSON, TOM AND MARC LIBARLE (eds.)   *The High School Revolutionaries*
V-279   SILBERMAN, CHARLES E.   *Crisis in Black and White*
V-681   SNOW, EDGAR   *Red China Today*

V-432   SPARROW, JOHN   *After the Assassination: A Positive Appraisal of the Warren Report*
V-222   SPENDER, STEPHEN   *The Year of the Young Rebels*
V-388   STAMPP, KENNETH   *The Era of Reconstruction 1865-1877*
V-253   STAMPP, KENNETH   *The Peculiar Institution*
V-454   STARR, PAUL AND IMMANUEL WALLERSTEIN (eds.)   *The University Crisis Reader: The Liberal University Under Attack,* Vol. I
V-455   STARR, PAUL AND IMMANUEL WALLERSTEIN (eds.)   *The University Crisis Reader: Confrontation and Counterattack,* Vol. II
V-613   STERNGLASS, ERNEST J.   *The Stillborn Future*
V-374   STILLMAN, E. AND W. PFAFF   *Power and Impotence*
V-439   STONE, I. F.   *In a Time of Torment*
V-547   STONE, I. F.   *The Haunted Fifties*
V-231   TANNENBAUM, FRANK   *Slave & Citizen: The Negro in the Americas*
V-312   TANNENBAUM, FRANK   *Ten Keys to Latin America*
V-322   THOMPSON, E. P.   *The Making of the English Working Class*
V-686   WALLACE, MICHAEL AND RICHARD HOFSTADTER (eds.)   *American Violence: A Documentary History*
V-206   WALLERSTEIN, IMMANUEL   *Africa: The Politics of Independence*
V-543   WALLERSTEIN, IMMANUEL   *Africa: The Politics of Unity*
V-454   WALLERSTEIN, IMMANUEL AND PAUL STARR (eds.)   *The University Crisis Reader: The Liberal University Under Attack,* Vol. I
V-455   WALLERSTEIN, IMMANUEL AND PAUL STARR (eds.)   *The University Crisis Reader: Confrontation and Counterattack,* Vol. II
V-145   WARREN, ROBERT PENN   *Segregation*
V-323   WARREN, ROBERT PENN   *Who Speaks for the Negro?*
V-405   WASSERMAN AND SWITZER   *The Random House Guide to Graduate Study in the Arts and Sciences*
V-249   WIEDNER, DONALD L.   *A History of Africa: South of the Sahara*
V-557   WEINSTEIN, JAMES   *Decline of Socialism in America 1912-1925*
V-585   WEINSTEIN, JAMES AND DAVID EAKINS (eds.)   *For a New America*
V-605   WILLIAMS, JOHN A. AND CHARLES HARRIS (eds.)   *Amistad 1*
V-660   WILLIAMS, JOHN A. AND CHARLES HARRIS (eds.)   *Amistad 2*
V-651   WILLIAMS, WILLIAM APPLEMAN   *The Roots of the Modern American Empire*
V-313   WILSON, EDMUND   *Apologies to the Iroquois*
V-208   WOODWARD, C. VANN   *Burden of Southern History*
V-545   WOOLF, S. J. (ed.)   *The Nature of Fascism*
V-495   YGLESIAS, JOSE   *In the Fist of the Revolution*
V-483   ZINN, HOWARD   *Disobedience and Democracy*

V-340   ADAMS, RUTH (ed.) *Contemporary China*

V-286   ARIES, PHILIPPE *Centuries of Childhood*

V-185   BARNETT, A. DOAK *Communist China and Asia*

V-620   BILLINGTON, JAMES H. *Icon and Axe: An Interpretive History of Russian Culture*

V-44   BRINTON, CRANE *The Anatomy of Revolution*

V-250   BURCKHARDT, C. J. *Richelieu: His Rise to Power*

V-391   CARR, E. H. *What Is History?*

V-628   CARTEY, WILFRED and MARTIN KILSON (eds.) *Africa Reader: Colonial Africa*, Vol. I

V-629   CARTEY, WILFRED and MARTIN KILSON (eds.) *Africa Reader: Independent Africa*, Vol. II

V-556   CASTRO, JOSUE de *Death in the Northeast: Poverty and Revolution in the Northeast of Brazil*

V-518   CHILDE, V. GORDON *The Dawn of European Civilization*

V-526   DEHIO, LUDWIG *The Precarious Balance*

V-746   DEUTSCHER, ISAAC *The Prophet Armed*

V-747   DEUTSCHER, ISAAC *The Prophet Unarmed*

V-748   DEUTSCHER, ISAAC *The Prophet Outcast*

V-471   DUVEAU, GEORGES *1848: The Making of A Revolution*

V-611   FONTAINE, ANDRE *History of the Cold War*, Vol. I

V-612   FONTAINE, ANDRE *History of the Cold War*, Vol. II

V-475   GAY, PETER *The Enlightenment: The Rise of Modern Paganism*

V-277   GAY, PETER *Voltaire's Politics*

V-685   HAMSIK, DUSAN *Writers Against Rulers*

V-114   HAUSER, ARNOLD *The Social History of Art* through V-117 (four volumes)

V-630   HOROWITZ, DAVID *Empire and Revolution*

V-201   HUGHES, H. STUART *Consciousness and Society*

V-514   HUNTINGTON, SAMUEL P. *The Soldier and the State*

V-550   JENKINS, ROMILLY *Byzantium: The Imperial Centuries A.D. 610-1071*

V-50   KELLY, AMY *Eleanor of Aquitaine and the Four Kings*

V-628   KILSON, MARTIN and WILFRED CARTEY (eds). *Africa Reader: Colonial Africa*, Vol. I

V-629   KILSON, MARTIN and WILFRED CARTEY (eds). *Africa Reader: Independent Africa*, Vol. II

V-728   KLYUCHEVSKY, V. *Peter the Great*

V-246   KNOWLES, DAVID *Evolution of Medieval Thought*

V-83   KRONENBERGER, LOUIS *Kings and Desperate Men*

V-215   LACOUTURE, JEAN *Ho Chi Minh*

V-522   LANGER, WILLIAM L. *European Alliances and Alignments*

V-364   LEFEBVRE, GEORGES *The Directory*

V-343   LEFEBVRE, GEORGES *The Thermidorians*

V-587   LEWIN, MOSHE *Lenin's Last Struggle*

V-474   LIFTON, ROBERT JAY *Revolutionary Immortality: Mao Tse-Tung and the Chinese Cultural Revolution*

V-487   LIFTON, ROBERT JAY *Death in Life: Survivors of Hiroshima*

V-533   LOCKWOOD, LEE *Castro's Cuba, Cuba's Fidel*

V-92   MATTINGLY, GARRETT *Catherine of Aragon*

V-689   OBSERVER, AN *Message from Moscow*

V-733   PARES, SIR BERNARD *The Fall of the Russian Monarchy*

V-525   PARES, SIR BERNARD *A History of Russia*

V-285   PARKES, HENRY B.   *Gods and Men*
V-719   REED, JOHN   *Ten Days That Shook the World*
V-176   SCHAPIRO, LEONARD   *The Government and Politics of the Soviet Union* (Revised Edition)
V-745   SCHAPIRO, LEONARD   *The Communist Party of the Soviet Union*
V-375   SCHURMANN, F. and O. SCHELL (eds.)   *The China Reader: Imperial China,* I
V-376   SCHURMANN, F. and O. SCHELL (eds.)   *The China Reader: Republican China,* II
V-377   SCHURMANN, F. and O. SCHELL (eds.)   *The China Reader: Communist China,* III
V-681   SNOW, EDGAR   *Red China Today*
V-312   TANNENBAUM, FRANK   *Ten Keys to Latin America*
V-322   THOMPSON, E. P.   *The Making of the English Working Class*
V-724   WALLACE, SIR DONALD MACKENZIE   *Russia: On the Eve of War and Revolution*
V-206   WALLERSTEIN, IMMANUEL   *Africa: The Politics of Independence*
V-298   WATTS, ALAN   *The Way of Zen*
V-557   WEINSTEIN, JAMES   *The Decline of Socialism in America 1912-1925*
V-106   WINSTON, RICHARD   *Charlemagne: From the Hammer to the Cross*
V-627   WOMACK, JOHN JR.   *Zapata and the Mexican Revolution*
V-81   WOOCK, ROGER R. and ARTHUR I. BLAUSTEIN (eds.)   *Man against Poverty: World War III*
V-486   WOOLF, S. J. (ed.)   *European Fascism*
V-545   WOOLF, S. J. (ed.)   *The Nature of Fascism*
V-495   YGLESIAS, JOSE   *In the Fist of Revolution: Life in a Cuban Country Town*

VINTAGE WORKS OF SCIENCE
AND PSYCHOLOGY

V-286   ARIES, PHILIPPE   *Centuries of Childhood*

V-292   BATES, MARSTON   *The Forest and the Sea*

V-129   BEVERIDGE, W. I. B.   *The Art of Scientific Investigation*

V-291   BIEBER, I. AND OTHERS   *Homosexuality*

V-532   BOTTOMORE, T. B.   *Critics of Society: Radical Thought in North America*

V-168   BRONOWSKI, J.   *The Common Sense of Society*

V-419   BROWN, NORMAN O.   *Love's Body*

V-338   CHURCH, JOSEPH   *Language and the Discovery of Reality*

V-410   CHURCH, JOSEPH (ed.)   *Three Babies: Biographies of Cognitive Development*

V-157   EISELEY, LOREN   *The Immense Journey*

V-390   ELLUL, JACQUES   *The Technological Society*

V-248   EVANS, JEAN   *Three Men*

V-413   FRANK, JEROME D.   *Sanity and Survival: Psychological Aspects of War and Peace*

V-132   FREUD, SIGMUND   *Leonardo da Vinci: A Study in Psychosexuality*

V-14   FREUD, SIGMUND   *Moses and Monotheism*

V-124   FREUD, SIGMUND   *Totem and Taboo*

V-493   FREUND, JULIEN   *The Sociology of Weber*

V-491   GANS, HERBERT J.   *The Levittowners*

V-195   GRODDECK, GEORG   *The Book of the It*

V-404   HELLER, WALTER (ed.)   *Perspectives on Economic Growth*

V-283   HENRY, JULES   *Culture Against Man*

V-521   HENRY, JULES   *Jungle People*

V-663   HERRIGEL, EUGEN   *Zen in the Art of Archery*

V-397   HERSKOVITS, MELVILLE J.   *The Human Factor in Changing Africa*

V-566   HURLEY, RODGER   *Poverty and Mental Retardation: A Causal Relationship*

V-268   JUNG, C. G.   *Memories, Dreams, Reflections*

V-636   JUNG, C. G.   *Analytical Psychology: Its Theory and Practice*

V-436   KAUFMANN, WALTER   *Nietzsche: Philosopher, Psychologist, Antichrist*

V-437   KAUFMANN, WALTER (trans.)   *The Will to Power*

V-337   KAUFMANN, WALTER (trans.)   *Beyond Good and Evil*, by Friedrich Nietzsche

V-369   KAUFMANN, WALTER (trans.)   *The Birth of Tragedy and The Case of Wagner*, by Friedrich Nietzsche

V-401   KAUFMANN, WALTER (trans.)   *On the Genealogy of Morals and Ecce Homo*, by Friedrich Nietzsche

V-210   KENYATTA, JOMO   *Facing Mount Kenya*

V-361   KOMAROVSKY, MIRRA   *Blue-Collar Marriage*

V-226   KROEBER, A. L. AND CLYDE KLUCKHOLN (eds.)   *Culture*

V-164   KUHN, THOMAS S.   *The Copernican Revolution*

V-492   LEFEBRVE, HENRI   *The Sociology of Marx*

V-426   LEKACHMAN, ROBERT   *The Age of Keynes*

V-105   LESLIE, CHARLES (ed.)   *Anthropology of Folk Religion*

V-280	LEWIS, OSCAR	*The Children of Sánchez*
V-421	LEWIS, OSCAR	*La Vida: A Puerto Rican Family in the Culture of Poverty—San Juan and New York*
V-370	LEWIS, OSCAR	*Pedro Martínez*
V-284	LEWIS, OSCAR	*Village Life in Northern India*
V-634	LEWIS, OSCAR	*A Death in the Sánchez Family*
V-487	LIFTON, ROBERT JAY	*Death in Life: Survivors of Hiroshima*
V-650	LIFTON, ROBERT JAY	*Boundaries*
V-690	LIPTON, ROBERT JAY	*History and Human Survival*
V-384	LINDESMITH, ALFRED	*The Addict and the Law*
V-76	LINTON, RALPH	*The Tree of Culture*
V-209	MARCUSE, HERBERT	*Eros and Civilization*
V-579	NISBET, ROBERT A.	*Tradition and Revolt*
V-639	OUSPENSKY, P. D.	*Tertium Organum*
V-672	OUSPENSKY, P. D.	*The Fourth Way*
V-558	PERLS, F. S.	*Ego, Hunger and Aggression: Beginning of Gestalt Therapy*
V-462	PIAGET, JEAN	*Six Psychological Studies*
V-528	PRESTHUS, ROBERT	*The Organizational Society*
V-70	RANK, OTTO	*The Myth of the Birth of the Hero* and Other Essays
V-99	REDLICH, FRITZ M.D. AND JUNE BINGHAM	*The Inside Story: Psychiatry and Everyday Life*
V-622	ROAZEN, PAUL	*Freud: Political and Social Thought*
V-682	ROBINSON, JOAN	*Freedom and Necessity*
V-395	ROKEACH, MILTON	*The Three Christs of Ypsilanti*
V-301	ROSS, NANCY WILSON (ed.)	*The World of Zen*
V-464	SARTRE, JEAN-PAUL	*Search for a Method*
V-647	SEXTON, PATRICIA CAYO	*The Feminized Male*
V-289	THOMAS, ELIZABETH MARSHALL	*The Harmless People*
V-310	THORP, EDWARD O.	*Beat the Dealer*
V-588	TIGER, LIONEL	*Men in Groups*
V-299	WATTS, ALAN	*The Joyous Cosmology: Adventures in the Chemistry of Consciousness*
V-468	WATTS, ALAN	*The Wisdom of Insecurity*
V-592	WATTS, ALAN	*Nature, Man, and Woman*
V-665	WATTS, ALAN	*Does it Matter?*
V-298	WATTS, ALAN	*The Way of Zen*

V-365   ALPEROVITZ, GAR   *Atomic Diplomacy*
V-604   BAILYN, BERNARD   *The Origins of American Politics*
V-334   BALTZELL, E. DIGBY   *The Protestant Establishment*
V-198   BARDOLPH, RICHARD   *The Negro Vanguard*
V-60   BECKER, CARL L.   *The Declaration of Independence*
V-494   BERNSTEIN, BARTON J. (ed.)   *Towards a New Past: Dissenting Essays in American History*
V-199   BERMAN, HAROLD J. (ed.)   *Talks on American Law*
V-211   BINKLEY, WILFRED E.   *President and Congress*
V-512   BLOCH, MARC   *The Historian's Craft*
V-513   BOORSTIN, DANIEL J.   *The Americans: The Colonial Experience*
V-358   BOORSTIN, DANIEL J.   *The Americans: The National Experience*
V-621   BOORSTIN, DANIEL J.   *The Decline Of Radicalism: Reflections on America Today*
V-44   BRINTON, CRANE   *The Anatomy of Revolution*
V-98   CASH, W. J.   *The Mind of the South*
V-311   CREMIN, LAWRENCE A.   *The Genius of American Education*
V-190   DONALD, DAVID   *Lincoln Reconsidered*
V-379   EMERSON, THOMAS I.   *Toward a General Theory of the First Amendment*
V-424   FOREIGN POLICY ASSOCIATION, EDITORS OF   *A Cartoon History of United States Foreign Policy Since World War I*
V-498   FRANKLIN, JOHN HOPE   *From Slavery to Freedom: History of Negro Americans*
V-368   FRIEDENBERG, EDGAR Z.   *Coming of Age in America*
V-264   FULBRIGHT, J. WILLIAM   *Old Myths and New Realities*
V-463   GAY, PETER   *A Loss of Mastery: Puritan Historians in Colonial America*
V-400   GENOVESE, EUGENE D.   *The Political Economy of Slavery*
V-676   GENOVESE, EUGENE D.   *The World the Slaveholders Made*
V-31   GOLDMAN, ERIC F.   *Rendezvous with Destiny*
V-183   GOLDMAN, ERIC F.   *The Crucial Decade—and After: America, 1945-1960*
V-95   HOFSTADTER, RICHARD   *The Age of Reform: From Bryan to F.D.R.*
V-9   HOFSTADTER, RICHARD   *The American Political Tradition*
V-317   HOFSTADTER, RICHARD   *Anti-Intellectualism in American Life*
V-385   HOFSTADTER, RICHARD   *The Paranoid Style in American Politics and Other Essays*
V-540   HOFSTADTER, RICHARD and CLARENCE L. VER STEEG (eds.)   *Great Issues in American History, From Settlement to Revolution, 1584-1776*
V-541   HOFSTADTER, RICHARD (ed.)   *Great Issues in American History, From the Revolution to the Civil War, 1765-1865*
V-542   HOFSTADTER, RICHARD (ed.)   *Great Issues in American History, From Reconstruction to the Present Day, 1864-1969*
V-591   HOFSTADTER, RICHARD   *Progressive Historians*
V-630   HOROWITZ, DAVID   *Empire and Revolution: A Radical Interpretation of Contemporary History*
V-514   HUNTINGTON, SAMUEL P.   *The Soldier and the State*
V-242   JAMES, C. L. R.   *The Black Jacobins*
V-527   JENSEN, MERRILL   *The New Nation*

V-623   KRADITOR, AILEEN S. *Means and Ends in American Abolitionism*

V-367   LASCH, CHRISTOPHER *The New Radicalism in America*

V-560   LASCH, CHRISTOPHER *The Agony of the American Left*

V-488   LYND, STAUGHTON *Intellectual Origins of American Radicalism*

V-502   MATTHEWS, DONALD R. *U. S. Senators and Their World*

V-552   MAYER, ARNO J. *Politics and Diplomacy of Peacemaking*

V-386   MCPHERSON, JAMES *The Negro's Civil War*

V-318   MERK, FREDERICK *Manifest Destiny and Mission in American History*

V-84   PARKES, HENRY B. *The American Experience*

V-371   ROSE, WILLIE LEE *Rehearsal for Reconstruction*

V-212   ROSSITER, CLINTON *Conservatism in America*

V-285   RUDOLPH, FREDERICK *The American College and University: A History*

V-394   SEABURY, PAUL *Power, Freedom and Diplomacy*

V-279   SILBERMAN, CHARLES E. *Crisis in Black and White*

V-52   SMITH, HENRY NASH *Virgin Land*

V-345   SMITH, PAGE *The Historian and History*

V-432   SPARROW, JOHN *After the Assassination: A Positive Appraisal of the Warren Report*

V-388   STAMPP, KENNETH M. *The Era of Reconstruction 1865-1877*

V-253   STAMPP, KENNETH M. *The Peculiar Institution*

V-110   TOCQUEVILLE, ALEXIS DE *Democracy in America*, Vol. I

V-111   TOCQUEVILLE, ALEXIS DE *Democracy in America*, Vol. II

V-103   TROLLOPE, MRS. FRANCES *Domestic Manners of the Americans*

V-516   ULAM, ADAM B. *The Unfinished Revolution*

V-540   VER STEEG, CLARENCE L. and RICHARD HOFSTADTER (eds.) *Great Issues in American History, 1584-1776*

V-265   WARREN, ROBERT PENN *The Legacy of the Civil War*

V-605   WILLIAMS, JOHN A. and CHARLES F. HARRIS (eds.) *Amistad 1*

V-660   WILLIAMS, JOHN A. and CHARLES F. HARRIS (eds.) *Amistad 2*

V-362   WILLIAMS, T. HARRY *Lincoln and His Generals*

V-208   WOODWARD, C. VANN *Burden of Southern History*

VINTAGE CRITICISM,
LITERATURE, MUSIC, AND ART

V-418 AUDEN, W. H. *The Dyer's Hand*

V-398 AUDEN, W. H. *The Enchâfed Flood*

V-269 BLOTNER, JOSEPH and FREDERICK GWYNN (eds.) *Faulkner at the University*

V-259 BUCKLEY, JEROME H. *The Victorian Temper*

V-51 BURKE, KENNETH *The Philosophy of Literary Form*

V-643 CARLISLE, OLGA *Poets on Streetcorners: Portraits of Fifteen Russian Poets*

V-569 CARTEY, WILFRED *Whispers from a Continent: The Literature of Contemporary Black Africa*

V-75 CAMUS, ALBERT *The Myth of Sisyphus and other Essays*

V-626 CAMUS, ALBERT *Lyrical and Critical Essays*

V-535 EISEN, JONATHAN *The Age of Rock: Sounds of the American Cultural Revolution*

V-655 EISEN, JONATHAN *The Age of Rock 2*

V-4 EINSTEIN, ALFRED *A Short History of Music*

V-632 ELLMAN, RICHARD (ed.) *The Artist as Critic: Critical Writings of Oscar Wilde*

V-13 GILBERT, STUART *James Joyce's Ulysses*

V-646 GILMAN, RICHARD *The Confusion of Realms*

V-363 GOLDWATER, ROBERT *Primitivism in Modern Art*, Revised Edition

V-114 HAUSER, ARNOLD *Social History of Art*, Vol. I

V-115 HAUSER, ARNOLD *Social History of Art*, Vol. II

V-116 HAUSER, ARNOLD *Social History of Art*, Vol. III

V-117 HAUSER, ARNOLD *Social History of Art*, Vol. IV

V-438 HELLER, ERICH *The Artist's Journey into the Interior and Other Essays*

V-213 HOWE, IRVING *William Faulkner: A Critical Study*

V-20 HYMAN, S. E. *The Armed Vision*

V-12 JARRELL, RANDALL *Poetry and the Age*

V-88 KERMAN, JOSEPH *Opera as Drama*

V-260 KERMODE, FRANK *The Romantic Image*

V-581 KRAMER, JANE *Allen Ginsberg in America*

V-452 KESSLE, GUN, photographs by, and JAN MYRDAL *Angkor*

V-83 KRONENBERGER, LOUIS *Kings and Desperate Men*

V-677 LESTER, JULIUS *The Seventh Son*, Vol. I

V-678 LESTER, JULIUS *The Seventh Son*, Vol. II

V-90 LEVIN, HARRY *The Power of Blackness: Hawthorne, Poe, Melville*

V-296 MacDONALD, DWIGHT *Against the American Grain*

V-55 MANN, THOMAS *Essays*

V-720 MIRSKY, D. S. *A History of Russian Literature*

V-344 MUCHNIC, HELEN *From Gorky to Pasternak*

V-452 MYRDAL, JAN and photographs by GUN KESSLE *Angkor*

V-118 NEWMAN, ERNEST *Great Operas*, Vol. I

V-119 NEWMAN, ERNEST *Great Operas*, Vol. II

V-24 RANSOM, JOHN CROWE *Poems and Essays*

V-108 SHAHN, BEN *The Shape of Content*

V-415 SHATTUCK, ROGER *The Banquet Years*, Revised

V-186 STEINER, GEORGE *Tolstoy or Dostoevsky*

V-278 STEVENS, WALLACE *The Necessary Angel*

V-39 STRAVINSKY, IGOR *The Poetics of Music*

V-100   SULLIVAN, J. W. N.   *Beethoven: His Spiritual Development*
V-243   SYPHER, WYLIE (ed.)   *Art History: An Anthology of Modern Criticism*
V-266   SYPHER, WYLIE   *Loss of the Self*
V-229   SYPHER, WYLIE   *Rococo to Cubism*
V-458   SYPHER, WYLIE   *Literature and Technology*
V-166   SZE, MAI-MAI   *The Way of Chinese Painting*
V-162   TILLYARD, E. M. W.   *The Elizabethan World Picture*
V-35   TINDALL, WILLIAM YORK   *Forces in Modern British Literature*
V-194   VALERY, PAUL   *The Art of Poetry*
V-347   WARREN, ROBERT PENN   *Selected Essays*
V-218   WILSON, EDMUND   *Classics & Commercials*
V-360   WIMSATT, W. and C. BROOKS   *Literary Criticism*
V-500   WIND, EDGAR   *Art and Anarchy*
V-546   YATES, FRANCES A.   *Giordano Bruno and the Hermetic Tradition*